# 𝒜 VIEW OF THE
# UNIVERSE

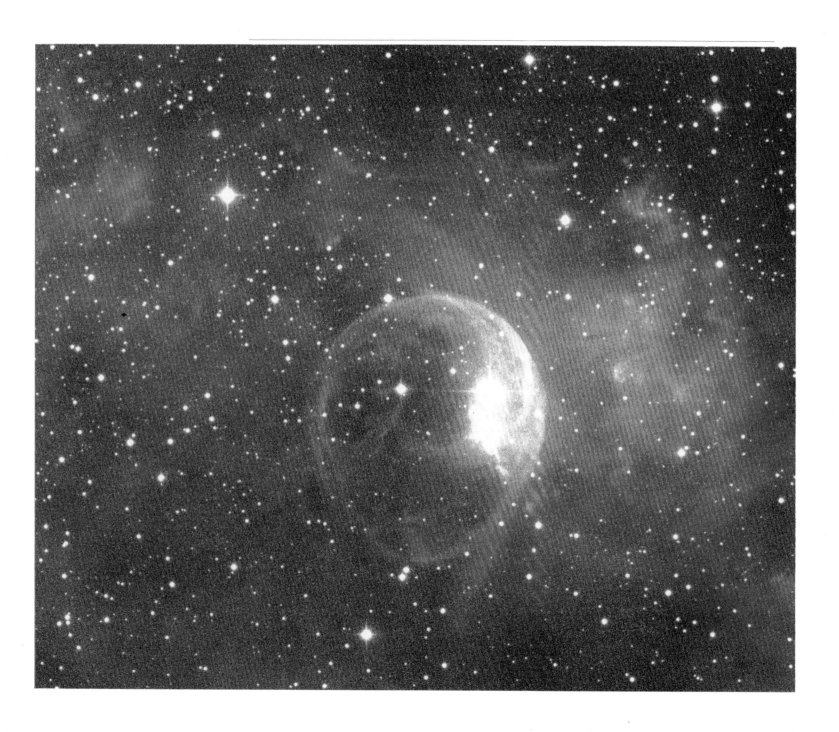

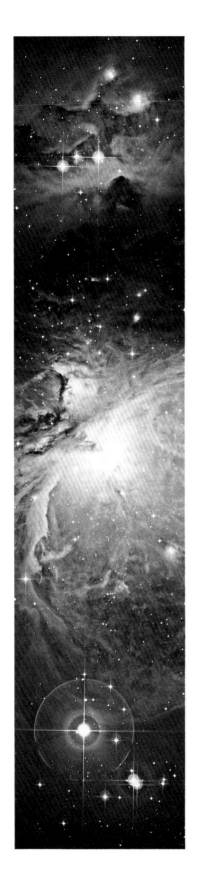

# *A* VIEW OF THE
# UNIVERSE

## DAVID MALIN

*Anglo-Australian Observatory*

## SKY PUBLISHING CORPORATION

*Cambridge, Massachusetts*

and

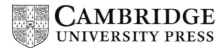 CAMBRIDGE
UNIVERSITY PRESS

Published by Sky Publishing Corporation
49 Bay State Road, Cambridge, Massachussetts 02138
and by the Press Syndicate of the University of Cambridge
The Pitt Building, Trumpington Street, Cambridge CB2 1RP
40 West 20th St, New York, NY 10011-4211, USA
10 Stamford Road, Oakleigh, Melbourne 3166, Australia

First published 1993

Printed in Great Britain at the University Press, Cambridge

*A catalogue record for this book is available from the British Library*

*Library of Congress cataloguing in publication data*

Malin, David,
A view of the universe / David Malin.
  p. cm.
  Includes index.
  ISBN 0 933346 66 2 (Sky Publishing Corporation)
  ISBN 0 521 44477 2 (Cambridge University Press)
  1. Astronomical photography.  2. Colour photography.  I. Title.
QB121.M34  1993
522'.63–dc20  92-46469 CIP

ISBN 0 933346 66 2 hardback (Sky Publishing Corporation)
ISBN 0 521 44477 2 hardback (Cambridge University Press)

TAG

FOR PHILLIPA

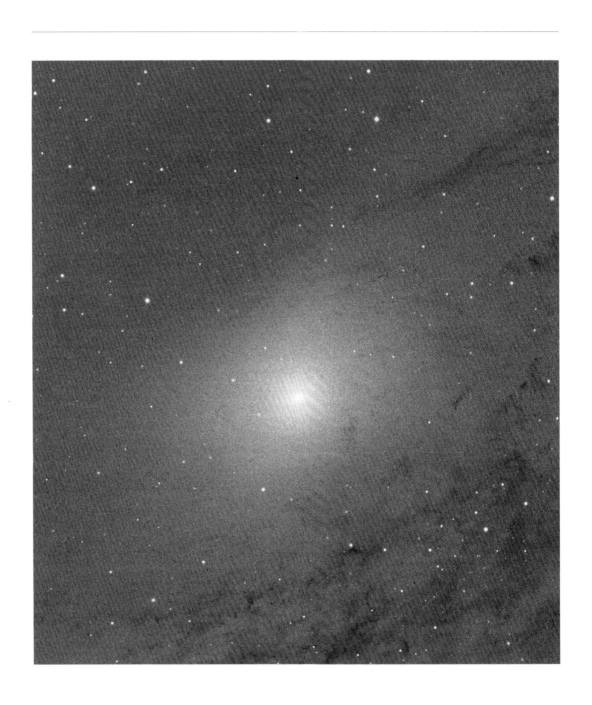

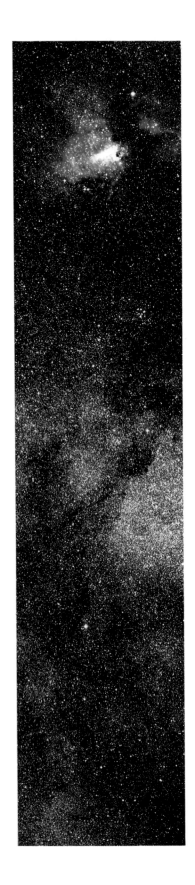

# CONTENTS

# INTRODUCTION

## NIGHT AND DAY

The difference between night and day is more than merely dark and light. For half our lives the most distant view we have is limited by the blueish haze of sunlight, scattered by the air we breathe. Only the Sun itself and occasionally a pale Moon hint that there may be something beyond the comfortable, familiar blue of the daytime sky. At night, when the Earth has turned its back to the Sun, its light no longer illuminates the air above. For a few hours we can see through the turbulent atmosphere to the Universe beyond. What we see with the unaided eye tells us very little. Only with the invention of the telescope did we begin to understand that this planet is not at the centre of all things. Only with the arrival of photography did we realise that there was no centre. This new view of the Universe continues to have a profound effect on the intellectual framework of modern society.

The daily cycle of the world regulates our clocks and calendars, and has set the internal rhythms of our bodies, governing the pattern of sleep and wakefulness. But nowadays many of us live in cities, insulated from the realities of nature by tall buildings, air-conditioning and a sky that is always bright, night and day. For many of us the horizon is but a distant memory, and a rainbow seen recently in Broadway produced a column of comment in the *New York Times*.

It is therefore not surprising that very few trouble to gaze upward. If they did they would be regarded as eccentric; from the glittering canyons of the city and the brightly lit suburbs there is little to see. The electric light and airborne pollution have conspired to hide the stars, but they are still there, begging an explanation, mute testimony to our origins and our home, embraced in the arms of a spiral galaxy.

I have been fortunate to have the opportunity to explore the sky at night with camera and telescope, and within the pages of this book is a pictorial record of one such exhilarating voyage of discovery that has lasted 18 years and still continues. The images are proof of the beauty and endless variety of nature. However, these photographs are not merely a record of an exploration, they are the essential tool of discovery, revealing more than eye and telescope together can see. Almost alone among those who work with visible light, the photographic astronomer is only rarely able to see what is being sought and is never able to admire it in full colour.

The photographic process is therefore a vital part of the search, and in my case was the entrée into the explorer's club. Much as a terrestrial explorer might emphasise the strength of his boat, or the challenges of navigation, the demanding photographic techniques used to create the pictures are not ignored here, mainly because they influence the pictures themselves and thus become part of the interpretation. Although peculiar photographic skills and large telescopes are needed to probe the more distant reaches covered on this journey, much of interest can be done with an ordinary camera and colour film. A little patience and imagination are required, as well as an escape from city lights, but the results are rewarding and revealing. Both approaches to photography in the dark are described in non-technical terms, and are linked to show how straightforward pictures of the stars made with everyday equipment can uncover important facts about them.

The essence of this book is the importance of colour in understanding astronomical objects and phenomena. Since the beginning of photography, astronomers had grown used to studying in black and white the painstakingly captured images of distant objects. Often their monochromes were themselves records of light that had been filtered into a narrow range of colours, sometimes into the three primary shades of blue, green and red. Though the colour information that they captured was vital for astronomical interpretation, the black and white images were usually compared side by side, often in the form of a negative, where bright stars are black points and wispy galaxies grey smudges. The colour information that these plates contained had to be absorbed gradually and with patient study; it never appeared as a colour picture outside the observer's imagination.

Despite their unpromising appearance, and the difficulty in understanding them, such black and white colour separations have the convenient property of containing all the information necessary to reconstruct a full-colour image of a scene. This was the basis of James Clerk Maxwell's famous 1861 demonstration of the feasibility of colour photography. Because the pictures are in the form of black and white negatives they can be manipulated and copied in the black and white domain to enhance the feeble or tone down the overbearing in a way that is not possible with colour film.

This allows the three colour separations, often enhanced, manipulated and improved, to be combined later into pictures that not only reveal the astonishing visual impact of colour in astronomical images, but contain otherwise hidden scientific information. Despite this roundabout way of making pictures, the aim throughout has been to portray the colours true to life, much as they would appear if the contrast and colour sensitivity of the eye or the brightness of the nebula had been increased by several orders of magnitude. The photographic processes that perform this transformation are introduced in Chapter 1, and a bibliography provides references to the original work.

From most of the black and whites, colour images have been created, selected and arranged to illuminate some facets of the many topics that engross researchers into the physical nature of the Universe beyond the solar system. Here and there, new information revealed in colour has some feature better illustrated in monochrome, and such pictures are scattered throughout the book, interspersed with a few black and whites of interesting and illuminating examples of objects for which no colour image is available.

However, colour is the key to understanding, and for the purpose of this book, where the colour images end, so does the exploration.

Each chapter has a brief introduction and the subsequent text is woven among the pictures to describe what they show and explain why it is important or interesting. While many of the photographs stand on their own as images to enjoy, the satisfaction of looking at photographs, even the most abstract forms, is enhanced if the subject matter is identified, so the pictures themselves carry captions: sometimes brief identifiers, sometimes most of the story.

Chapter 1 is used to explain how a chemist from the north of England came to take photographs with one of the finest optical telescopes in the world, located on a mountain-top in outback Australia, and it outlines how some of the photographic techniques used to create the pictures in these pages evolved. In Chapter 2 is a heartfelt tribute to photography, which has transformed the way in which we view the world around us and brought to us the previously unsuspected, invisible worlds of the microscope and telescope. Despite the recent rise of the electronic detector, photography's long legacy has extended our vision and provided most of the basic information we have about the Universe beyond the surface of the Earth. The theme of perception is continued in Chapter 3 with a series of images that explore the daytime sky with the eye of a night-time astronomer. After sunset we show how the results of simple photography of the stars with an ordinary camera can be linked to the knowledge that we live inside a spiral galaxy. This comes partly from understanding the colours of the stars.

Within that galaxy, we find the formation of the stars takes place in spectacular nebulae which are laced with material that was once inside an earlier generation of stars that are now long gone. Many of the pictures here are of the beautiful colours and forms produced by the birth and death of the stars. We continue our search for colour and enlightenment beyond the Milky Way to other galaxies, near and far. With increasing distance the colours are less strident and the signs harder to read as we try to make sense of ever feebler signals from more distant parts of the Universe.

In the end, towards the edge of the Universe, at least those parts of it accessible to terrestrial 4 metre telescopes, we must acknowledge that we are unable to register the most distant objects in colour. Like the colour response of the eye itself, colour photography has its limits, mainly because on the largest scale the Universe itself becomes undifferentiated, distant galaxies blending one into the other. But the often subtle hues we can reveal on our long journey has expanded our view of that part of the Universe that we can see, and given it an important new dimension, colour.

## NAMES AND MAGNITUDES

Throughout this book of pictures, the galaxies and nebulae that appear are referred to in a variety of ways and some objects are referred to by several different designations, so naturally there may be a little confusion, especially since, like most astronomers, I have not attempted to be systematic. While a completely systematic procedure for naming astronomical objects might be thought to be desirable, and is often discussed, such rigidity would mask some of the romance of astronomy and deny its long and colourful heritage.

Some of the names of objects point to the constellation where they are found. The constellation names themselves are very old and reflect the importance of the stars to ancient cultures. For example, we discuss the Great Nebula in Orion and the star clouds of Sagittarius, but there is more than one nebula in Orion and the star clouds of the Milky Way extend well beyond Sagittarius, so while we acknowledge that these names are imprecise they have a certain poetry that catalogue names lack. They also recall the rich and fascinating history of astronomy. The knowledge that the Galaxy was composed of vast numbers of faint stars had to await the invention of the telescope in 1609, but the names of the constellations that straddle the Milky Way are thousands of years old. They reflect the fascinating interplay between the human mind and the long search for meaning among the random pattern of the stars. On another matter of nomenclature, the first letter of 'our' Galaxy is conventionally capitalised to distinguish it from galaxies in general.

Star names too are often derived from languages and cultures long dead. Betelgeuse is the name of the bright yellow star in Orion. According to Richard Hinckley Allen in his wonderful *Star Names, Their Lore and Meaning*, its name is derived from ancient Arabic 'Ibt al Jauzah', and means armpit of the central one. The word Orion itself has perhaps come to us from an ancient Akkadian tongue, which had its origins 4000 years ago in what is now Iraq.

We also refer to objects by their Messier number, sometimes abbreviated, as in M42, which is another name for the Orion nebula. Charles Messier was an eighteenth-century French comet hunter for whom misty-looking objects such as nebulae, galaxies and globular clusters were an observational inconvenience. To avoid confusion with the equally fuzzy-looking but mobile comets that he sought, he made a list of the positions and descriptions of about 100 fixed, extended objects that he found could be confused with them. Since his list includes many of the brightest non-stellar objects in the sky, Messier's catalogue has been referred to for 200 years and 'M' numbers still appear in the astronomical literature.

The most common prefixes used to identify objects in this book are the NGC numbers. 'NGC' refers to the *New General Catalogue*, the first modern 'all sky' listing. It contains almost 8000 assorted objects that were found by searching the sky by eye and telescope, mainly by William Herschel in the northern hemisphere, and his son John, who scanned the south in the 1830s. It has been supplemented by the *Index Catalogue*, two lists of smaller, often fainter things that the Herschels missed, so the combined IC and NGC list contains nearly 13 000 entries. The Orion nebula is NGC 1976. The *New General Catalogue* was compiled by John Dreyer in 1888, so 'new' in this context is relative.

It soon became clear that photography was capable of finding so many new objects that cataloguing them all was impossible, so most modern lists are of single species, planetary nebulae, for example, or galaxies. Since in this book we concentrate mostly on the bigger and brighter objects, Messier and NGC numbers predominate in our index. But some objects are not bright; Barnard 86, for example, is a dark cloud, number 86 in Edward Emerson Barnard's list of such things. Malin-1 is a faint but very massive galaxy, the first and so far only entry in an embarrassingly short list.

There are also some oddities in the text with peculiar, non-systematic names such as Centaurus A. This is the designation given by radio astronomers to the strongest radio source in that constellation. It was soon found to coincide with the bright galaxy NGC 5128, but both names have stuck and both are used by people who talk about radio sources and optical galaxies. Where an object is referred to by more than one name, they all appear with their equivalents in the index.

Astronomers have adopted the vague word 'magnitude', which to most people means size, importance or quantity, and given it a precise meaning. A second magnitude star is, by definition, 2.512 times fainter than a first magnitude star, and a difference of 5 magnitudes is a factor of exactly 100 in brightness. The bigger the number, the fainter the star, so that first magnitude stars are bright and easily seen, sixth magnitude stars are on the threshold of naked eye visibility and the faintest stars on our photographs are twenty-fourth magnitude or even fainter.

# ACKNOWLEDGEMENTS

It is important first to recognise that the photographs in this book would not have been made without the ongoing commitment of the Eastman Kodak Company to the production of a range of unique spectroscopic photographic emulsions. These specialised materials are produced to order and coated on glass, almost exclusively for the small but demanding scientific photography market. These plates have been used to prepare most of the colour pictures reproduced here that were taken with telescopes. This is one area of colour photography where Kodak has no commercial competition.

This book would also not have been possible without help from the many friends and colleagues at the Anglo-Australian Observatory who have helped with prime focus photography on the Anglo-Australian Telescope over the years. Special thanks are due to the staff of the United Kingdom Schmidt Telescope, who took many of the black and white plates that became the colour pictures identified as UKS images in the long list reproduced as the Appendix. These individuals are identified at the end of that table and are warmly thanked for their willingness to experiment and for the consistent and unsurpassed quality of their plates.

Belated thanks are due to Joe Wampler, the antithesis of a photographic astronomer, who, as first Director of the Anglo-Australian Observatory, encouraged me to look at ways of making colour photographs with the telescope. Don Morton, his successor, ensured that there was always a little telescope time in our Service Photography programme for public relations photography. The present Director of the Anglo-Australian Observatory, Russell Cannon, has been just as supportive, as has the Australian Time Assignment Committee, who since 1990 have provided several nights specifically for colour photography. I am also grateful to Francisco Sanchez, Director of the Instituto de Astrofisica de Canarias and his staff, especially Campbell Warden, for facilitating access to the Isaac Newton Telescope in October 1991, and to Derek Jones for making it all work for five hectic nights.

Thanks are also due to Paul Murdin, whose enthusiasm at the sight of the first three-colour pictures in 1976 encouraged me to make more; to David Allen, always helpful and stimulating, and a source of scientific rectitude; and to Keith and Sue Tritton, who helped to make observing at Siding Spring such a pleasure. They arranged for the first three-colour plates to be taken on the UK Schmidt Telescope on my behalf in 1978 and later.

This book has grown from an article I was asked to prepare for the Indian Academy of Science's *Current Science* journal. That in turn came from a series of lectures I gave in India in 1989–90 at the invitation of the Academy with the support of the Third World Academy of Sciences. Most of the talks were given in Bangalore, and I am grateful to Professor V. Radhakrishnan for making that period at the Raman Research Institute stimulating in so many ways, and to Professor S. Ramaseshan for insisting that I write up my talks in a manner sufficiently coherent for publication. Paul Murdin's comments on an early draft of this book improved it enormously.

I must also acknowledge the long-standing cooperation of KolorKraft in Sydney, who have done most of the processing and proofing of my difficult negatives for many years, and especially Ian Wall, now with Independent Display Services in Sydney, who has taken so much interest in the work. Bruce Scott has, with infinite care and patience, produced most of the print film transparencies that appear here.

Finally, it is a pleasure to thank the Eastman Kodak Company again, this time for their generous support in the preparation of this book, and especially for help with the creation of the many large-format transparencies that have been reproduced so well in these pages.

As this book was in the final stages of production, it became increasingly evident that Eastman Kodak were about to cease production of the special plates that were used to make the colour pictures, so the photographs around which this book was built may be the last of their kind.

# 1 FROM MICROSCOPE TO TELESCOPE

## INTRODUCTION

I often give 'popular' talks on astronomy and photography to clubs and societies of various kinds. They can be groups of amateur astronomers or photographers, high school students, Rotarians, even veterans of 25 years with IBM and other companies. The talks are usually a string of pictures tied together by loose themes and linked with a few anecdotes, rather like this book. There always seems to be an audience somewhere that is keen to see pictures of the night sky, and sometimes astonishing numbers of people turn up; in 1990 a public talk in Sydney Town Hall drew 1400 people, and in 1988, 600 Canarians filled a small theatre in La Laguna to hear a talk via simultaneous translation. This amazing feat was accomplished by Campbell Warden, the bi-lingual (Scots–Spanish) secretary of Francisco Sanchez, Director of the Instituto de Astro-fisica de Canarias. The translation was so effective that the audience even laughed at the jokes! That was the occasion that Paul Murdin and I spent the after-noon cleaning a huge, discoloured screen that had not seen the light of a projector for decades.

As you might imagine, such activities are great fun, and a totally unexpected conse-quence of becoming involved in astronomical photography. One of the many pleasures that accompany these occasions is the stream of questions after the talks and sometimes these continue for as long as the lecture that preceded them. The questions are enjoyable because they are unpredictable and dissolve away the barrier between lecturer and audi-ence, turning a monologue into a dialogue or discussion. It is always amazing (and very satisfying) that what I see as a fairly straightforward presentation can trigger so many different and stimulating responses. I suppose the kind of people who would trouble to attend such a talk are just those who have enquiring minds. Along with the usual ques-tions about life on other planets, the origins of UFOs and what film to use is the one that I normally try to evade. It often takes the form of asking how I became involved with astronomy, or wondering what qualifications are necessary to be an astronomical pho-tographer.

I avoid answering it because there is no short response I can give that would be in any way useful, nor is it possible to leave the real answer, that it was largely a matter of chance and good fortune, hanging there without further explanation.

1

## A CHEMICAL CAREER

I was interested in science from the time I discovered books, but I was born as an only child into an environment that was hardly touched by technology. For the first 10 years or so I lived in a house that had neither electricity nor gas and where the only source of cold water was an ancient brass tap over a solid sandstone sink. Cooking was on an open fire or on a Primus kerosene burner, and every Monday morning the laundry was boiled in a large cast-iron pot and squeezed dry through the huge, button-crushing wooden rollers of a hand-turned mangle.

The stone house was on the outskirts of the village of Summerseat, and it overlooked the valley of the River Irwell as it wound between Ramsbottom and Bury, in what was then industrial Lancashire. The view was surprisingly green towards the hundreds of distant factory chimneys I could count from the bedroom window. The only scientific instruments I recall in those years were a box camera, a pair of opera glasses, a wind-up gramophone and an alarm clock. These were progressively dismantled over the years, and I was astonished to find that the box camera could be persuaded to produce an image of the world that was both upside-down and in full colour. Even better was the objective lens of the opera glasses, which could also be used to start fires by focusing sunlight. None of these activities were particularly popular with the owners of these valuables.

No doubt it was with much relief that at the age of five I was sent to Walmersley Primary School when it was still in its pre-baby-boom 1834 sandstone building. There I was lucky to fall under the guidance of the headmaster, Leon Fell, the first 'educated' person I recall meeting, who eventually ensured that I passed the 11-plus examination that sent me to Bury High School. It was a good school in 1951, but my time there was mostly wasted, largely by my own indolence, but also because many (but by no means all) of the teachers were both old and of the old school, unaware of the fascination that the technology of the post-war era had for youngsters. But chemistry was made interesting by Mr Gibson, and it satisfied the urge to experiment that I seemed to have been born with.

When I eventually left the High School, voluntarily, but too soon to complete my education, I was again extremely fortunate to be offered a kind of laboratory assistant's apprenticeship in the Pesticides Laboratory of the Manchester branch of what was then the Geigy Company, a large, Swiss-based specialist chemical manufacturer. I still do not know why I was selected, an ignorant, very nervous 15 year old, devoid of experience or qualifications. But I was, and it was a turning point in my life. I began work there in February 1957.

As a condition of employment I was expected to attend technical college for one full day and three nights a week to obtain what amounted to a science degree with chemistry, physics and mathematics components. This process was in the end to take seven years, but much more important was the stimulating atmosphere of the laboratory in which I worked under the caring eye of George Lombard, with Michael Vessey and Donald Lea.

Though a small laboratory concerned with the testing and formulation of DDT,

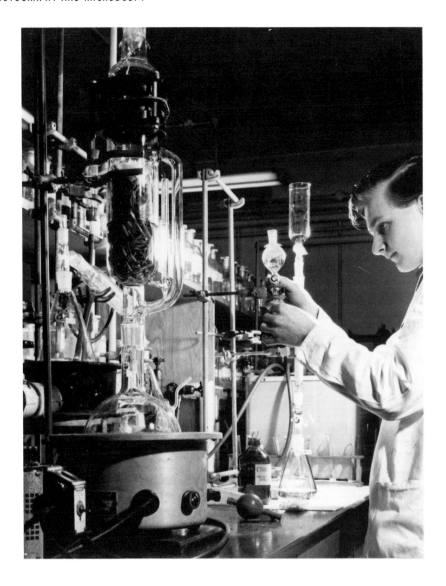

*Fig. 1.1.* Absorption chromatography and Soxhlet extraction were familiar tools of the chemist's trade when David Malin began his chemical career.

weed-killer and other pesticides, it was for me a kind of university, with other laboratories and other people nearby working on an enormous range of interesting topics, many of which I was eventually to become involved in myself. I suddenly discovered a new world of music, politics, books, films, foreign travel and endless stimulating conversation, which I now take for granted. And all the rooms had electricity and gas.

## PHOTOGRAPHY AND MICROSCOPY

During this period I also discovered photography. I was loaned a camera and used it on a visit to Paris, arriving home with a dozen rolls of 120-size black and white film and no money. I was encouraged by my colleagues to compound a developer from the basic ingredients in the lab and proceeded to process the films in the Company darkroom, an out-of-the-way cubby-hole, which until then had been used to house what seemed to me to be an antique microscope and an equally old and little-used enlarger.

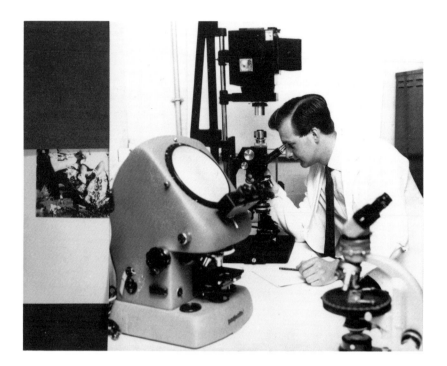

*Fig. 1.2.* Showing off a selection of the optical instruments used in the Microscopy and Technical Photography Unit for various kinds of chemical microscopy.

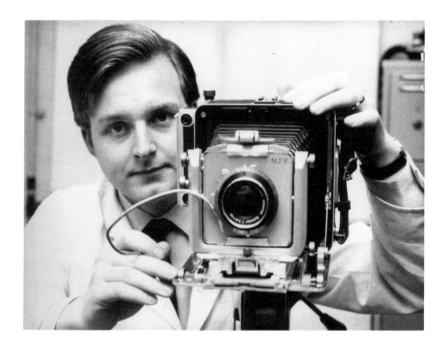

*Fig. 1.3.* The Microscopy Unit also had a photographic role, and large-format cameras were frequently used.

Faced with a large number of negatives, indifferently processed and exposed by guess-work, the novice printer set to work at weekends and evenings to make some respectable prints. I still recall the thrill of seeing the shape of the Eiffel Tower and the white domes of the Sacré Coeur magically appear in the developer, and was soon producing respectable prints on war-surplus photographic paper. Like many others before me, I discovered that I greatly enjoyed re-creating captured images of the world in the dark-

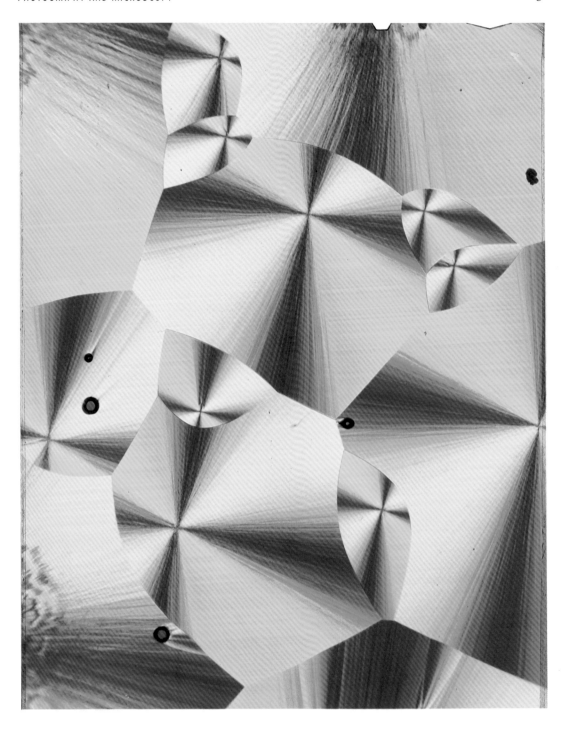

*Fig. 1.4.* An optical photo-micrograph of DDT spherulitic crystals seen through the polarising microscope.

room, and quickly bought my first camera, a well-used British device called an Agi-flex. It took twelve 6 × 6 centimetre negatives on 120 film, a format I still use.

This unscheduled darkroom activity, which inevitably spilled over into Company time, eventually came to the attention of management. Fortunately, the Geigy Company was a very enlightened organisation, and far from discouraging me, my new-found photographic skills were exploited. I was asked to use the darkroom's dusty microscope to

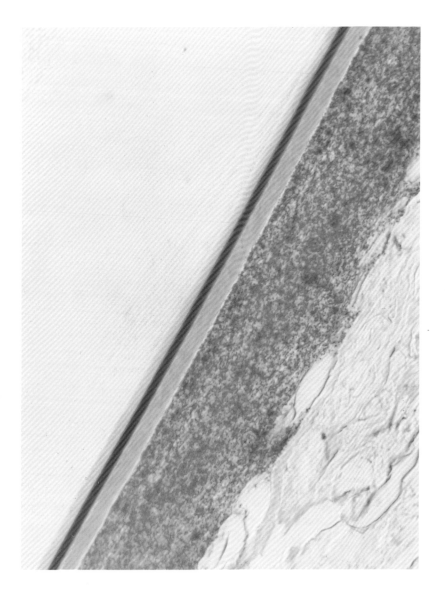

*Fig. 1.5.* A cross-section of an exposed and processed photographic colour print shows the separate yellow, magenta and cyan layers on a thick white pigment base coated over a rougher layer of paper fibres.

take some micrographs of a product whose solubility depended on its crystal properties. I assumed at the time they thought that anyone who could take photographs and who was keen enough to work in a stuffy closet must be willing to use the equipment it housed. I now see it as inspired man-management. It was also another stroke of luck.

The microscope, long neglected in a dusty corner, was an unending joy for the experimenter. Far from being a decorative antique, it was an old but little-used Watson Holophot, which was a very high quality research system, a kind of vertical optical bench on which everything was adjustable, quite unlike most of today's factory-aligned microscopes. Using this system effectively and to its full capability was a major challenge in which I became thoroughly immersed. I certainly felt thrown in at the deep end.

The microscope had accessories for phase contrast, polarisation, oil immersion, ultraviolet fluorescence, dark field, metallographic illumination and much else, everything that the serious microscopist could want or imagine. Very soon all these techniques

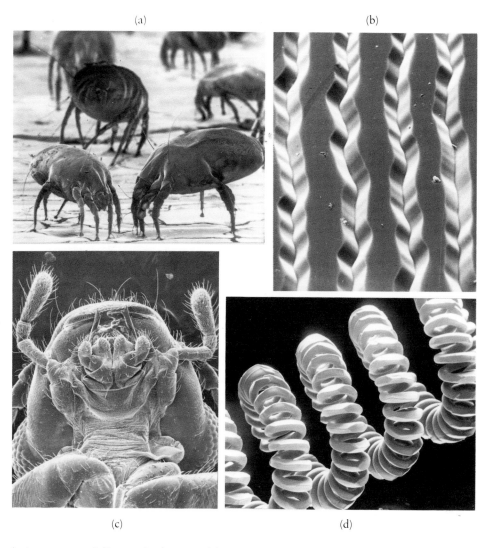

*Fig. 1.6.* A selection of pictures made with the scanning electron microscope: (a) house dust mite colony, (b) grooves on a stereo LP record, (c) an insect surrendering to the electron beam, (d) coiled coil lamp filament.

were being successfully applied to problems involving many of the Company's technical activities, often far removed from pesticides. It was not long before I was busy enough to start my own one-man lab, offering specialised microscopy services to other divisions within the rapidly expanding Geigy group.

The key ingredient in all of this was photography. It captured images of the tiny world of the microscope slide, an environment where surface tension and Brownian motion dominate and gravity is not important. It revealed the curious optical properties of fibres and the mechanisms of crystal growth. Everyday materials such as textiles, paint, paper and plastics took on a new appearance, which in turn led to a new understanding of how dyes, pigments, inks and lubricants did their work. And all this in a company that made such materials and set great store on understanding their manufacture, application and use.

The pictures that I took were used to explain this curious microcosm to other scientists, and gradually microscopy became an important part of many investigations, the micrographs illuminating hidden facets of the chemical processes involved in the research and development of existing products and the discovery of new ones. The small section that I ran soon became bigger as the company grew, merged with another, and became CIBA-Geigy, moving to Heald Green in Cheshire. The venerable Holophot was supplemented by other equipment as more complex investigations were undertaken and the value of applying physical methods to the solution of chemical problems became increasingly apparent. Eventually, we bought a scanning electron microscope, one of the first in the north of England.

This device was the perfect picture-producing machine. It created photographs that even management could understand, and revealed a whole new world of surfaces that were difficult to explore with a microscope that only used visible light. Once again, the photographs and their interpretation were the key to its success, and several examples are included here. Finally, I acquired an X-ray diffraction set, which is a kind of microscope for revealing facts about the ordered world of the crystal lattice, and even here photography was still used for recording some diffraction patterns, though most work was done with electronic detectors. This pattern was to be repeated in a subsequent career.

## UNSHARP MASKING

All of these techniques used photography in one way or another to convey their information and it was natural for me to explore ways of improving this channel of communication. I experimented with different developers and films and techniques for manipulating images. I was especially interested in ways of increasing photographic contrast to enhance photo-micrographs and often used high contrast films for this, long before it became fashionable. These difficult images were often impossible to print in any conventional way, so that sometimes I would use a copying technique that I had first encountered in the literature on printing processes while I was working on pigments for printing inks.

This was unsharp masking, a simple way of increasing the apparent sharpness of an image while avoiding excessive contrast. It works because photographic negatives and positives are complementary: superimpose one on the other and they cancel out. But if the positive is made unsharp, or blurred, it will only cancel the unsharp information in the negative, thus allowing the enhancement of fine detail without increasing contrast overall. It sounds complicated but is simple enough in practice and was soon to be very useful in an entirely different application.

Working in an environment where experimentation was encouraged I had become used to using photography in many non-standard ways and was increasingly fascinated by the physics and chemistry of the photographic process itself. It is still miraculous to me that the energy contained in a few photons of light can persuade a carefully formulated sea of electrons (a developer) to transform a crystal of silver bromide suspended in a clear gelatin layer into a lump of metallic silver (a photographic grain). I also enjoyed studying the history of the subject, and am still surprised that photography was discovered at all. The evolution and exploitation of the early colour processes is also especially revealing about how discovery and industry, art and science interact.

By this time I was well established, recognised as an expert chemical microscopist and a source of strong photographs for all kinds of purposes. I also ran a small department and thought I deserved Head of Lab status, but the then Director of Research did not agree, so I started to look for something else to do. As was entirely normal for someone in their early 30s, I now had a wife, three children and a mortgage, so some caution was necessary. But I suddenly realised that I had been with the same company for almost 18 years, so it was time for a change.

## OFF TO AUSTRALIA

My wife was not any ordinary woman. She was Australian, and we were happily married with a young family and settled in a country cottage in rural Cheshire. Without that connection it is unlikely that I would have looked twice at the advertisement in the back pages of *Nature,* which sought someone to establish the photographic laboratories of the Anglo-Australian Telescope (AAT). Without telling her, I applied for the job and made it to the first interview, woefully ignorant about astronomical photography. It was only during the interview that I began to realise what an enormous opportunity was on offer. I was therefore delighted and very surprised when, several months later, I was invited to a second interview, this time with Joe Wampler, then Director of the Observatory and a very laid-back and friendly Californian. By now I knew that I wanted the job, and so did my wife, but Joe did not ask me much about photography, rather he seemed more concerned about how the besuited Englishman before him would settle in the much less formal academic, scientific and social environment of Australia.

I shared his concern, but none the less we arrived in Sydney in August 1975, after I had visited the Royal Greenwich Observatory and the European Southern Observatory photolabs, then in Geneva, desperate to learn anything that would help me to start a fresh career in a (to me) new science in a recently established observatory on the other side of the world. I arrived to find my darkrooms, in the Sydney suburb of Epping, erected but unfinished in a temporary pre-fabricated building. I had an empty, windowless $7 \times 7$ metre concrete room which at this stage had neither water nor electricity (nor gas).

At the same time, an assistant photographer who was stationed at the telescope, 350 kilometres away at Siding Spring Observatory, had just resigned, so there was much to do there as well. But what to do first? Apart from the basic problems of plumbing and wiring there were processing lines to be set up, tray rockers to be designed and made, safelights, enlargers and general equipment to be ordered, and some basic sensitometry and densitometry to be established before experiments on hypersensitising could begin. The list was very long, and the problem was made more complex because I did not really know exactly how I wanted to do many of these things. But I was not working in a vacuum, and new colleagues, many of whom were to become my friends, offered plenty of advice and guidance.

The staff at the UK Schmidt Telescope were especially supportive at this time. They had already established a hypersensitising system at their telescope at Siding Spring. This method used a long-term soak (several weeks) in nitrogen to flush oxygen and water

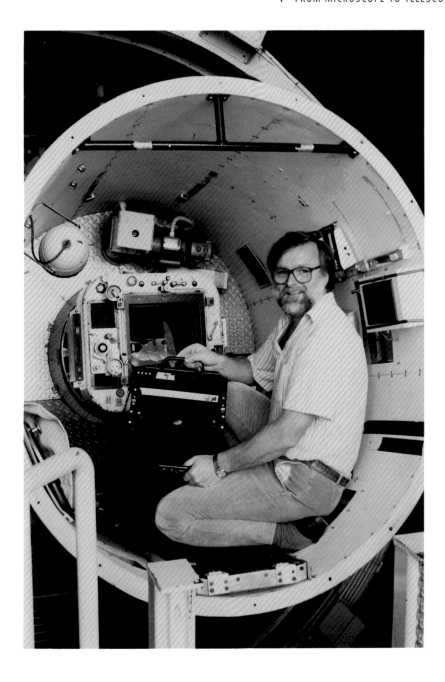

*Fig. 1.7.* A recent picture of David Malin in the prime focus cage of the Anglo-Australian Telescope.

from the photographic emulsion, thus improving its sensitivity to faint light. They were also just starting to use hydrogen, which improves sensitivity further, and they made their system available to me for preparing AAT plates. This was an enormous bonus, because it removed the necessity to duplicate their effort. We have used the Schmidt system for hydrogen treatment of AAT plates ever since.

But it was not just the specialised business of astronomical photography that I was bemused by. The Anglo-Australian Observatory was already producing exciting new science, and the place was alive with unfamiliar astronomical talk and ideas. It was impossible not to be swept along by the wave of enthusiasm generated by the first members of

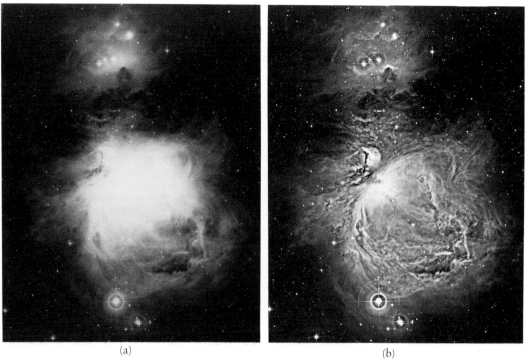

(a)                    (b)

*Fig. 1.8.* An early example of unsharp masking, here applied to the Orion nebula: (a) is a direct print, using the original UK Schmidt plate as a normal negative, while in (b) the print was made after the plate had been copied with an unsharp mask to reveal an astonishing amount of structure within the nebula.

the scientific staff as they exploited the wonderful opportunities presented by the finest and most sophisticated large telescope in the world, newly commissioned and sitting beneath the largely unexplored southern sky.

Although the astronomers were well aware of the potential of the telescope, Joe Wampler was very keen for us to have some colour pictures of astronomical objects to add to the scientific output, and to announce to the wider world that the AAT was up and running. We discussed at length the various options, including colour separation techniques using a dye transfer process, but I could not imagine investing the amount of time necessary to become skilled at that very demanding process for quite some time. Other possible methods required ways of superimposing multiple images on glass plates that seemed insurmountable at the time. In view of this and all the other pressing problems, as an interim measure I took a few images on 8 × 10 inch (20.3 × 25.4 cm) sheets of colour film stuck to glass plates because the AAT's prime focus camera was not (and is still not) equipped for film in any size. Among the first of these colour pictures was the spiral galaxy M83 (Fig. 1.15). Though we were well pleased with it at the time and the picture was widely published, I have made much better images since, and many of these are published elsewhere in this book.

## PHOTOGRAPHIC AMPLIFICATION AND ADDITION

Gradually, as the list of things to do became shorter, and equipment arrived and was installed, opportunities arose for more creative photographic work. An early project that I became involved with was illustrating a book with Paul Murdin and David Allen that was to appear as *Catalogue of the Universe* (Cambridge University Press, 1979). This was an education in itself and was an excellent incentive to discover some of the less

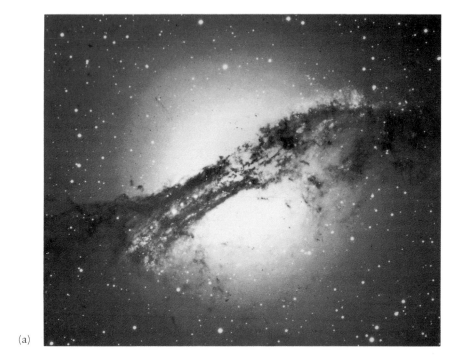

(a)

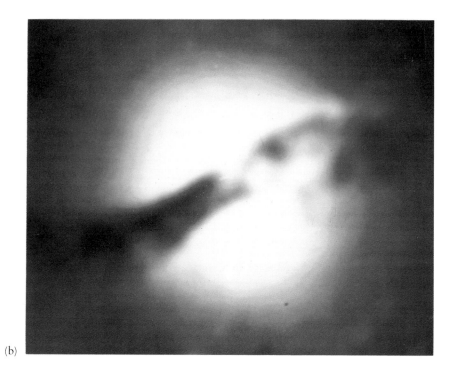

(b)

well-frequented parts of the extensive historical and scientific literature of astronomy. It was here that I published some of the examples of my early efforts at unsharp masking astronomical images. At some stage in the work we needed an eye-catching picture of an elliptical galaxy, but ellipticals are not especially interesting and in any case I did not

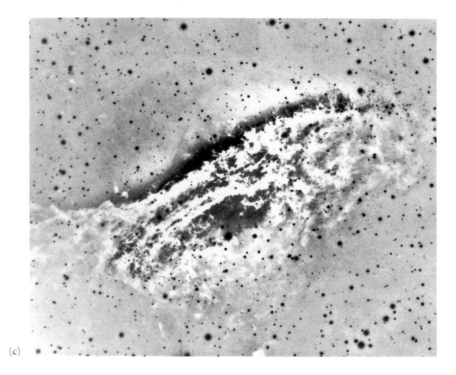

*Fig. 1.9.* Unsharp masking Centaurus A. A print made from the original AAT plate produces the familiar image of the galaxy (a). A deliberately blurred, positive film copy (b) contains no fine detail. By superimposing the blurred positive on the original negative most of the coarse detail in the original is cancelled, leaving the fine detail (c) previously hidden in the original negative.

(c)

have many very deep plates, except for one UK Schmidt plate of the Virgo cluster of galaxies.

In an attempt to make the Virgo ellipticals look bigger and more interesting I copied a part of the plate using high-contrast film. I was obliged to use the bench-top diffuse-light contact copier that I had installed as a temporary measure until I could construct a fancier point-light tunnel like the one I had seen at the European Southern Observatory. By chance I underexposed the first contrasty copy and was very surprised to see a feature apparently sticking out of the galaxy when I inspected the very weak-looking positive. Printing this on contrasty paper showed a large, faint extension to the galaxy, and an even fainter series of arcs or shells clearly centred on the galaxy, which was NGC 4552 (M89).

This was the first of a series of discoveries made with the technique that I called 'photographic amplification', and both the technique and the discovery of the M89 protrusion were written up as papers for *Nature*, again with invaluable help from colleagues in Epping. Most of the early photographic work in Epping was done with UK Schmidt plates because they were deep, uniform and of excellent quality, but many of the resulting publications were joint efforts with astronomers from the AAT who were better aware of the astronomical implications of what I was finding than I was at that time.

As more UK Schmidt plates became available it was possible to make photographically amplified copies of the same object from more than one plate. I made a simple superimposition frame that I could use to combine these separate positive copies into a single image. Combining or adding many images has the effect of enhancing the weak 'signal' from the object of interest (which appears on every plate) against the grainy

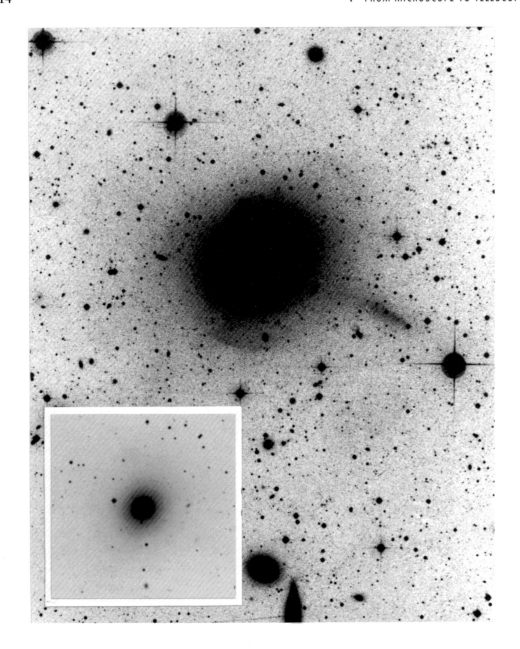

*Fig. 1.10.* One of the first apparently normal galaxies found to be peculiar by using photographic amplification was M89 in Virgo. The curious extension was discovered while making illustrations for a previous book and is probably a galaxy in the process of being disrupted by M89, which also shows a few other, shell or arc-like structures found at the same time. Inset is a 'normal' photograph of the galaxy, printed to the same scale.

'noise' from the photographic emulsion (which differs from plate to plate). This technique is important in revealing ever fainter objects.

Another technique based on accurate superimposition involves subtraction of images. Rather in the way that unsharp masking photographically removes an unwanted component of an object, everything (except the noise) in two identical photographs can be cancelled if a negative of one is precisely overlaid by a positive of the other. But if one of the images is slightly different, that difference is accentuated by the cancellation. This variation on the unsharp masking theme was combined with photographic amplification to explore the light echo from supernova 1987A. Precision registration of images is also the key to astronomical colour photography.

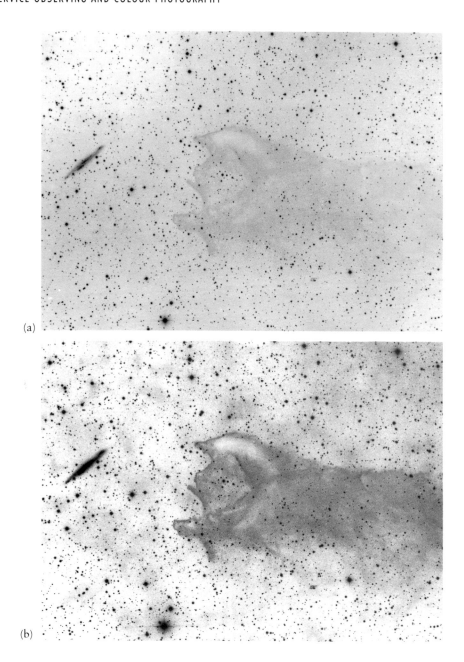

*Fig. 1.11.* Photographic amplification applied to three-colour photography: (a) shows the visual appearance of a deep AAT plate of the blue light exposure used for the first-ever colour picture of a cometary globule. After photographic amplification (b) much more faint detail is seen and the enhanced image can be incorporated into the colour picture reproduced as Fig. 5.27.

Both photographic amplification and unsharp masking are contact copying methods that produce a positive copy, usually on film, as their first derivative. It thus seemed natural to use a positive-working process to make colour pictures; after all, that is how the world's first colour picture was made in 1861. But how to obtain plates that contained the colour information in the first place?

## SERVICE OBSERVING AND COLOUR PHOTOGRAPHY

Time on a big telescope is usually allocated to well-defined projects, which are assessed by a time allocation committee (TAC) composed of other astronomers. It is a

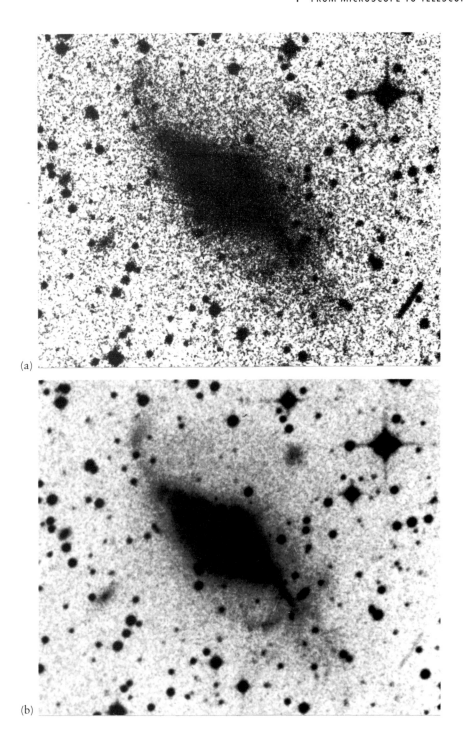

(a)

(b)

*Fig. 1.12.* The effect of multiple-image superimposition. A single, greatly enlarged and photographically amplified image of a peculiar galaxy is dominated by the grain 'noise' of the plate in (a). If eight such plates are added together, the grain noise varies from plate to plate and tends to cancel out, while the galaxy and stars are present on all the plates and are thus reinforced, producing a much 'cleaner' image (b). The galaxy is NGC 4672 in Centaurus.

peer review system that works reasonably well, but on a telescope like the AAT, which serves both the British and Australian communities, over-subscription is the norm. Quite often, there are only enough nights available in the 6-month allocation period to accommodate 20% or so of applications, so TACs have to be ruthless. As a consequence there is never any 'free' telescope time. Even commissioning time for new instruments and engineering time for fault finding is severely rationed. However, though it is usually only

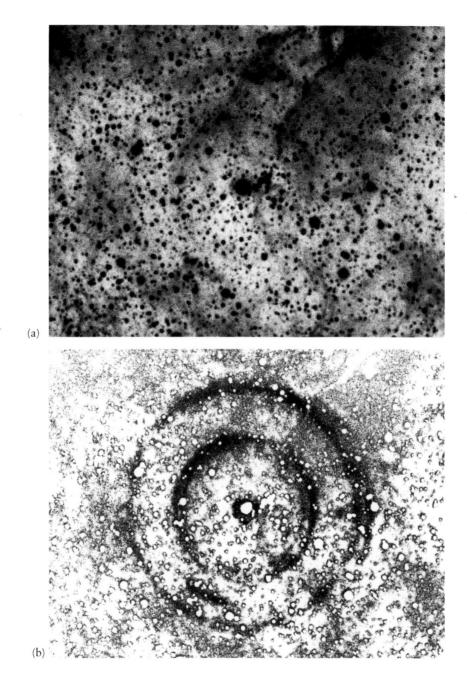

*Fig. 1.13.* The faint double circles of the light echo of supernova 1987A are just visible against the tangled nebulosity in the Large Magellanic Cloud in (a). This image was subtracted from another of the same part of the sky taken *before* the supernova appeared. The stars and nebulosity are common to both pictures and largely cancel, leaving the light echo as the dominant feature (b). The SN1987A light echo is discussed further in Chapter 5.

practical to schedule the telescope in full nights, some projects need only a few hours for a scientifically useful result, so if there are several such requests for similar kinds of observations, the Observatory bundles them all together into a single night. On these occasions the Observatory staff use the telescope on behalf of other astronomers as a kind of community service.

Photography lends itself very well to this mode of operation and service photography was begun by Don Morton shortly after he became Director of the AAT in 1976. I thought this an excellent scheme until I realised that I was going to be the observer, and

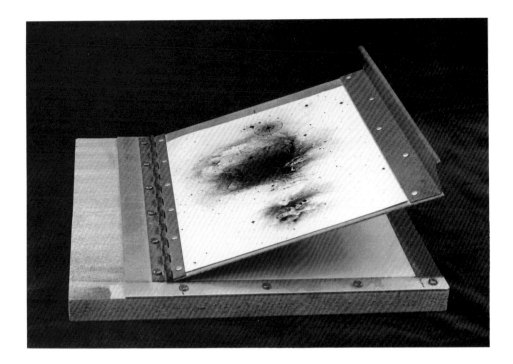

*Fig. 1.14.* A home-made superimposition frame is a simple device for use under the enlarger. A print stuck to the (normally flat) light-tight hinged lid acts as a positional reference for images to be combined on unexposed film or paper beneath the lid.

since I had never actually used the AAT (or any other telescope) before I was terrified of breaking it. Fortunately, Don appointed Louise Webster, as she was then, as support astronomer and it was she who taught me to use the telescope efficiently and how to prepare a programme for a night's observing. These are not trivial matters if best use is to be made of a night on a large instrument, especially since we usually had no more than six or eight nights a year, of which on average a third would be cloudy.

As a reward for running the programme, Don suggested informally that we might take one plate for ourselves during any one run. Though Louise and her many successors took very few plates for themselves, I began to fit plates for colour photography into the observing session, usually working well into twilight at both ends of the night. There are many ways that careful planning will save a few minutes here and there during the night, and it was a matter of pride that we could be collecting photons for well over 90% of the usable night on most occasions. Very few other modes of observation match that. However, even with the most careful organisation it is not always possible to end a planned 90 minute deep exposure scheduled for service observing exactly on twilight, so there were often a few precious minutes here and there.

Most of the plates for colour photography were taken in the well-established *B* (blue), *V* (green-yellow) and *R* (red) pass-bands that astronomers have used for photographic photometry for many years. This has many advantages, among them the fact that plate-filter combinations were readily available, and that many matched *B* and *V* pairs already existed, taken for purposes unconnected with colour photography. An exposure in red light might be all that was needed to complete a three-colour set for a new colour picture, and these could be done largely in twilight. It was in this way, and by using exposures taken for other projects, that negatives for the well-known

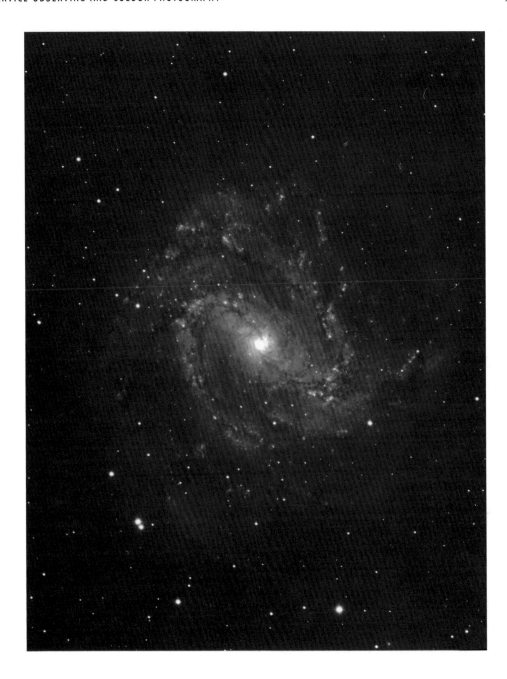

*Fig. 1.15.* One of the first AAT colour pictures made on colour film was of NGC 5236 (M83). Compare this with Fig. 4.10, a more recent picture made from three black and white plates.

series of colour pictures (AAT 7–54 in the Appendix) were obtained between 1976 and 1987.

Once processed, positive copies of these plates were made, often using unsharp masking or photographic amplification (or both) to control contrast or enhance faint detail. In many cases this produced pictures with astonishing colours where little colour was visible on photographs taken with colour film, as in the beautiful Rho Ophiuchi region. This was first photographed with colour film on the UK Schmidt Telescope, with lots of fiddling in the dark to tape 8 × 10 inch (20.3 × 25.4 cm) sheets of colour negative film somewhere near the centre of the 14 × 14 (35.6 × 35.6 cm) inch glass plates that the

*Fig. 1.16.* These two pictures are of the same part of the sky, and show the faint nebulae around Rho Ophiuchi and Antares. They were made with the UK Schmidt Telescope and show the difference between using ordinary colour film (left) and black and white (right) to make colour pictures.

telescope takes. The result was interesting, but not exciting. When the same field was re-photographed on three astronomical emulsions, stunning colours could be extracted from the black and whites to yield a picture that has been widely published. Occasionally, this kind of work in itself led to some scientific discovery, such as the Herbig–Haro objects found on the red-light plate of the Horsehead nebula (Fig. 7.2), which I worked on with Jeremy Walsh.

Once in the form of colour separation positives, the images could be combined onto colour negative film or positive-working colour paper to create a new colour picture that no-one had seen before. That, too, was often revealing, and I hope the excitement of that emerges from the pages of this book.

The slow accumulation of plates for AAT colour pictures stopped abruptly in February 1987, when a supernova appeared in the Large Magellanic Cloud (LMC). All service time and Director's time vanished as resources were diverted to this once-in-a-life-

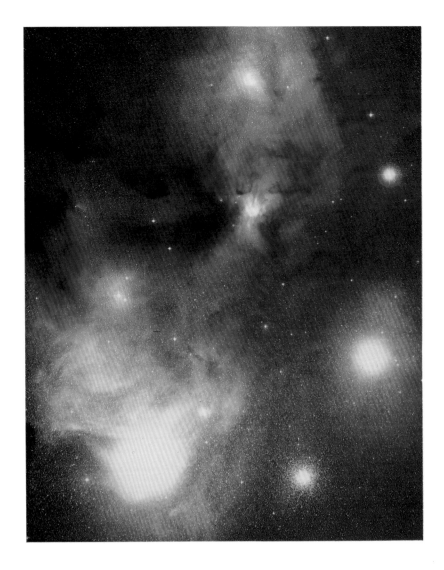

time event. Because of the vision of the Director, Russell Cannon, and the efforts of many staff members, the Anglo-Australian Observatory has the most complete (and still continuing) series of observations of the supernova. Now, however, I am delighted to say the Australian TAC has recently awarded several nights over the last two years specifically for colour photography, so the latest AAT colour pictures in these pages are Australian, though we are happy to share them with British readers.

The pleasure in these pictures comes because they communicate the beauty of the natural world to a wide audience and convey complex astrophysical concepts in a painless and even enjoyable way. They have certainly been an education to me. Unlike the microscope, however, these images are not visible in the eyepiece of the telescope. At best, even the most colourful gaseous nebulae seem to be little more than faint grey luminous smudges of light, almost indistinguishable from galaxies of stars. My role has been to give these fascinating images colour and, I hope, a greater meaning. References to the processes used to prepare the pictures in this book are given in the annotated Bibliography.

In the next chapter we consider how the limitations of human vision have influenced our view of the night sky and how photography has widened humankind's horizons, in 100 years completely changing our perspective of the Universe that we inhabit.

# 2 THE RISE OF ASTRONOMY AND PHOTOGRAPHY

## THE LIMITS OF DARKNESS

The limits that distance and darkness impose on human senses makes the night sky unique in our experience of the wider world. Though our field of view might be filled with points of light we can learn nothing of their size or shape or substance by simply looking. And although vision is less than perfect, it is all we have, for our other senses are of no use whatsoever. We cannot hear the stars, nor smell nor taste them, nor can we reach out and touch them to find out if they are near or far, hot or cold, friendly or alien. Indeed we can barely see them at all. In the struggle to collect enough of their light to render the stars visible, the eye loses its sense of colour and we notice only pallid hues even among the brightest constellations.

For this reason alone, the spectacle of a dark, star-filled night fills us with a sense of wonder, a need to know what it is we see. To make the unknown seem less threatening, our ancestors linked the random pattern of the stars into images of familiar creatures and protective gods that were important to their cultures. Given a rich tapestry of remembered patterns, it was no doubt soon noticed that superimposed on their nightly journey from east to west was another, more gradual drift, slowly changing with the seasons, yet repeating year by year. As well as the annual cycle of the stars there were the more erratic movements of the planets and the ever-changing face of the Moon. The predictable night-time events were punctuated by rare but dramatic occasions when the Sun vanished briefly by day or the Moon turned the colour of burnished copper as the bodies eclipsed one another.

Once the constancy of this pattern was understood and predictable, the objects in the sky could be relied upon as clock, calendar and compass, and the stars and planets appeared to be there for the convenience of humankind, who was so obviously at the centre of this wondrous system. Whereas elaborate 'world-systems' could be constructed to explain their movement, little could be learned of the Sun, the Moon or the stars themselves, or of the stars that wandered, the planets.

The Milky Way, too, was a complete mystery, and was to remain so until the inquisitive eye of Galileo Galilei peered into the winter sky of 1609 through a telescope he had made himself. Though his telescope was primitive and its images imperfect, Galileo was the first to see and sketch the mountains on the Moon. He noted the hour by

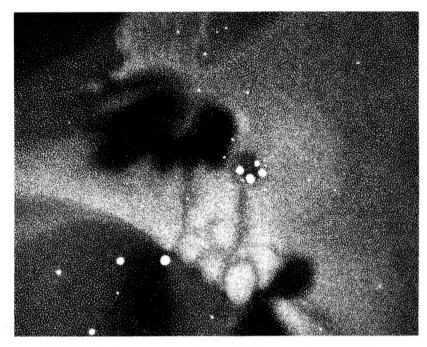

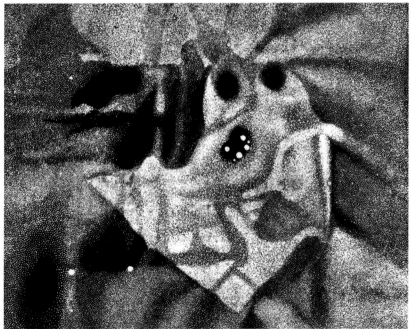

*Fig. 2.1* Upper: Orion nebula, Plate VIII in Sir John Herschel's *Results of astronomical observations at the Cape of Good Hope*, Smith Elder and Co., London, 1847. Lower: Orion nebula in 1868, from Lord Ross' *The scientific papers of the Earl of Rosse*, published by Sir Charles Parsons, London, 1926. Both illustrations have been reversed from the original negative sketches and printed to about the same scale. Though many of the main features are common to both drawings, there are many differences in rendition of detail. These pictures should be compared with Fig. 6.11, a modern colour picture of the brightest part of the Orion nebula.

hour, night by night meanderings of the satellites in orbit around the giant planet Jupiter and perhaps most important of all, discovered that the broad span of the Milky Way was made of countless numbers of faint stars.

Suddenly the Universe seemed more puzzling than anyone had imagined and immediately these new discoveries challenged the accepted wisdom. Armed with ever larger telescopes, and with minds alerted by the observations of Galileo, the Renaissance astronomers began to push back the barriers of ignorance and superstition that had prevented the growth of knowledge for over 1000 years. As bigger and better telescopes were turned to the sky, the Universe seemed suddenly very large, and ever more wonderful and mysterious sights were seen. In many cases, the mystery was deepened because even experienced observers could seldom agree on exactly what they saw. The only way to record what was seen was by a delicately pencilled sketch, scribbled in the dark and redrawn from memory by the light of day in a stronger but no more certain hand.

No doubt crude telescopes and the blurring of their images by the incessant trembling of the Earth's atmosphere added some degree of illusion to what was there, but the main problem was human. Of what the eye saw, the mind could remember only a fraction, and what the mind remembered, the hand could trace only imperfectly.

Within these limitations astronomy flourished. Humankind was forever displaced as the pivot of the Universe as the positions of stars and orbits of planets were determined with ever greater precision. But little sense could be made of the large numbers of faint, fuzzy objects that were dimly perceived between them. And though the stars themselves appeared brighter and much more numerous in a telescope, they were still examined with the only light-sensitive device then known, the human eye.

The subjective nature of drawing can be seen in the two sketches of the brightest of the diffuse nebulae, that in Orion, made by experienced observers and published in 1847 and 1868. Though they agree about the principal stars reproduced in each of the drawings, there is little correlation between structures in the luminous material between them. Little wonder that these early explorers thought the nebula restless, ever- changing, as they compared their drawings year by year. We now know that the nebulae are indeed turbulent, but the changes are so subtle and the objects so distant that they can hardly ever be detected by even the sharpest human eye over the span of a human life.

## THE CAPTURED IMAGE

But change of another kind was in the air. In 1839 the Frenchman Louis Daguerre announced the first practical system of photography. He had found a way of recording the pattern of light formed by a lens on an unlikely combination of a silvered copper plate, which had been treated with iodine fumes, and mercury vapour. Almost simultaneously, William Fox Talbot in England announced his negative–positive process, the forerunner of today's photography. These early processes, though quite different in detail, had several similarities. The most obvious of these to modern practitioners was that they were extremely inconvenient to use, requiring a mobile laboratory to prepare the plates, which by current standards were extremely insensitive to light. Pictures taken in even the brightest sunshine took so long to expose that even the most patient sitter froze into the poses that are so familiar to collectors of early portraits. The stiff-looking impressions we have of our Victorian forebears are not necessarily a reflection of their lives and times.

Thus, apart from the Sun itself, the Moon and a few of the brighter stars, the photography of barely visible astronomical objects was for long regarded as quite impossible. This remained the case for almost 50 years, but by the early 1880s technical improvements had made the photographic plate much more sensitive and easier to use, and astronomers slowly realised that many of the practical problems that had prevented photographic exploration of the night sky might be overcome. In 1880 the brightest of the diffuse nebulae, that in Orion, was photographed by Henry Draper in the United States and, with more success, by Andrew Ainslee Common in England three years later.

Though Draper's picture was the first to show the nebulosity, Ainslee Common's effort was technically superior and was at once recognised as a major achievement, though not because of the unexpected extent of the nebula. Close inspection of the plate revealed something even more astonishing. Hidden in the tangled wisps of nebulosity were images of stars so faint that they had never been seen before, even by the keenest-eyed astronomer. And many had looked, for the Orion nebula had been a great mystery for 200 years.

Ainslee Common's picture turned out to be much more than simply the first accurate record of the structure of what was soon to be known as a gaseous nebula; it was also the first to reveal objects whose existence was quite unsuspected. The photographic plate had not only fulfilled the need for an accurate sketch-pad, faithfully recording what was visible, it also turned out to be a superb detector of the unseen as well. Like the invention of the telescope 280 years before, the ability of the photographic plate to gather faint light for long periods was to revolutionise humankind's view of the Universe. At long last the human eye had a rival, and was rapidly overtaken. Astronomical photography had arrived.

Before long the adventurous spirits at every major observatory had tried their hand at photography with the telescope and discovered, often to their surprise, that it was a difficult art to master. In part this was because many of the telescopes of the time were designed to be looked through, and the photographic images they produced were poor. Also, most instruments could not follow the steady drift of stars across the sky accurately enough to avoid smearing out the images during the necessarily long exposures. Whereas an unsteady telescope drive mechanism was merely inconvenient to the visual observer, it was an absolute disaster for the new breed of photographic astronomers. It was in this area that Ainslee Common, who was a professional engineer, made some important advances. He also used a reflecting telescope rather than one that collected light through glass lenses, and this too gave him an advantage.

The gradual realisation that an hour's photography could reveal more than 200 years of patient looking triggered a new wave of telescope construction to exploit its unique ability to build a strong image from feeble light. These new telescopes were designed not for peering at the stars but for photographing them. They were installed in domes so that they were not shaken by the wind, and equipped with elaborate drive systems so that they could follow a chosen object for many hours without blurring the image that was slowly being collected by the light-sensitive layer.

Though the first of the new telescopes were refractors, with large lenses to focus the light, soon they were replaced by reflectors, whose size was not constrained by the mechanical properties of glass. By the beginning of this century the joint exploration of the Universe by telescope and photographic plate was well underway and there began a series of discoveries which have completely changed ideas about the size, age and structure of the Universe, and thus our perception of humankind's place within it.

While the arrival of photography has revolutionised astronomy, astronomy in turn has had its effect on photographic science. It was soon realised that photographing the night sky placed unusually severe demands on conventional photographic materials, and manufacturers began to produce small, experimental batches of plates specially designed for astronomical use, a tradition that has been continued by the Eastman Kodak Company to the present time.

In turn, astronomers have always been adept at making the very best use of the meagre amount of light that their telescopes collected by resorting to bizarre and sometimes dangerous ways of making their plates ever more sensitive. At one time they held their breath as they soaked their plates in choking ammonia solutions or suspended them in highly toxic mercury vapour just before use. Nowadays, the custom is to bake the plates in a moderate oven for an hour or two, followed by a soak in hydrogen gas before they are exposed at the telescope. The reasons for using these exotic and potentially dangerous hypersensitisation processes are buried in the literature of photographic science but there is no doubt as to their effectiveness. After gas treatment, the latest Kodak emulsions are among the most sensitive photographic materials ever made. Most of the three-colour photographs reproduced here have been made from black and white plates that have had their speed increased in this way.

Once the plates have been baked, exposed and carefully processed one by one, there comes the business of making sense of the myriads of tiny black dots and smudges that are revealed on the negative. There are as many ways of doing this as there are problems to be solved. Simply looking at the image of a distant galaxy may tell experienced astronomers all they want to know. More often, the plate will be used to measure the precise position of an optical object perhaps first detected at radio or X-ray wavelengths. The plate may also be compared with another, taken years before, to reveal changes in brightness or position, or compared with yet another made with light of a different colour to take the temperatures of thousands of stars and galaxies.

Unlike most ordinary negatives, the images recorded in the silver particles within the photographic layer are precious data that can be turned into numbers and searched for facts, just as one might extract data stored in the magnetic particles of a computer's disk. The same concepts of signal-to-noise and storage capacity that are familiar in any electronic channel of communication can therefore be applied to the silver image. But the plates are primarily photographs, and the information they have gathered can be used to make pictures that anyone can enjoy, without the aid of a computer.

Because of the peculiar nature of astronomical photography and the special, high contrast materials that are needed to record the faintest objects, equally special methods are often required to make useful pictures from them. My versions of these processes are discussed in Chapter 1, and the Bibliography lists sources of further technical details.

Although photographic plates are still used for specialised purposes and are essential for wide-angle surveys of the sky, the task of recording the well-travelled photons of astronomy is increasingly being undertaken by electronic sensors such as the charge coupled device (CCD), which is similar to the detectors used in compact TV cameras. The CCD's silicon chip is more sensitive than the silver grains in the photographic plate and its data can be read directly by a computer, but because it has lower resolution and a much smaller useful area the pictures it produces seem neither as spectacular nor as beautiful as those obtained by conventional photography. In addition, despite its rapid and widespread adoption, the CCD has not yet managed to reproduce the astonishing wave of astronomical discovery that marked the serious introduction of photography to astronomy a little over 100 years ago. It is more likely to be the CCD's close ally, the computer, that with hindsight will be seen to be the agent of astronomical revolution in the last quarter of the twentieth century. Already, the computer is extracting amazing images from feeble signals that would be unusable without sophisticated enhancement. The reconstruction of the stunning Voyager pictures of the outer members of the solar system and the restoration of the blurred images from the Hubble Space Telescope are conspicuous recent examples that come to mind.

It may be that the pictures in the following pages are the final flourish of a century of photographic achievement, which has transformed the science of astronomy and with it our view of the Universe, but somehow I doubt it. Before long, a new breed of young astronomers will make the astounding discovery that there exists a detector of high sensitivity, astonishing resolution and unlimited area that is compact, sturdy and cheap, and capable of detecting the faintest objects. The images it makes can even be viewed without resort to a computer, and they contain vast amounts of valuable information that can be displayed and enjoyed in full colour.

In the meantime, we begin our exploration of the night sky at the other end of the scale, with photographs that anyone can make. It is not always necessary to use a big telescope or advanced photographic techniques to enjoy the colours of nature. In the next chapter I will look at the sky and the stars using an ordinary camera and everyday colour film.

# 3 THINGS TO SEE AND DO IN THE DARK

## THE DAYTIME SKY

Since the beginnings of consciousness, daily life on planet Earth has been divided into two by our experience of space and time, the space in which we live and the regular cycle of light and dark. Only since our species began to inhabit cities and lose touch with nature have these perceptions shrunk to mean four walls and an alarm clock. Since most of the readers of this book will be city or suburban dwellers and will think of time and space in this way, I had better explain what I mean.

Our sense of space is usually defined by the divide that separates the Earth, on which we stand, from the sky, which we cannot reach. Our sense of the passage of time is similarly punctuated by the dark of night and warmth of day. Before technology made bright artificial light and electronic entertainment widely available, the most visually stimulating of the waking hours were between sunrise and sunset, a time when the sky seems largely devoid of interest. Naturally enough, most people spent the daytime with their eyes lowered to earthly matters. But the pursuit of astronomy is different in many ways from everyday life, and so most of this book is concerned with looking upwards in the hours of darkness. Since we are now examining a day and night from the astronomical point of view, perhaps it is useful to begin by considering why it is dark at night and light by day.

This is no place to elaborate on Olber's paradox, which asks why, if there exists an infinity of stars, the night-time sky is not filled with their light. This is not meant to be a philosophical book, despite its somewhat questioning opening paragraphs, and such topics can be explored in the publications in the reading list in the Bibliography. For philosophical purposes, the sky is dark at night because the Universe is young. For practical purposes it is dark at night because the rotation of the Earth has carried the nearest star below the horizon.

By day, the atoms and molecules of the air scatter some of the sunlight passing through the atmosphere. This scattering, named after Lord Rayleigh, who gave the first complete explanation, is caused by particles that are very small compared with the wavelength of light. What is distinctive about Rayleigh scattering is that the short wavelengths of light, (that part of the spectrum that we see as blue) are scattered more readily than longer (red) wavelengths. This is why the daytime sky is such a wonderfully uniform azure.

This intangible, luminous dome is visible proof of the otherwise more or less transparent atmosphere that protects the fragile life on Earth from the manifold hazards of the

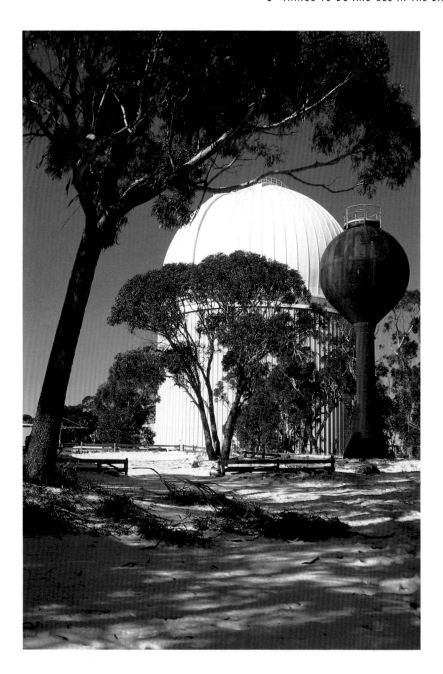

*Fig. 3.1.* The brilliant blue of the Australian sky is reflected in rare winter snow in the shadow of gum trees at Siding Spring Observatory, while the AAT dome and a rusty water tower are seen in direct sunlight.

rest of the Universe. It absorbs harmful ultraviolet rays and energetic atomic particles from the Sun and censors out all but the largest of the numerous solid particles that enter it day and night at terrifying speed.

Like some over-protective womb, guarding the life within, the daytime sky hides from us more of the realities of outer space than is strictly necessary for our survival. The soft blue haze completely blocks our view of the Universe beyond the solar system. Only the Sun and Moon, Venus, Mercury and (rarely) Jupiter are bright enough to be seen through it. By day, these nearby bodies are the only visible signs that the Earth and its pale blue sky is not all that there is.

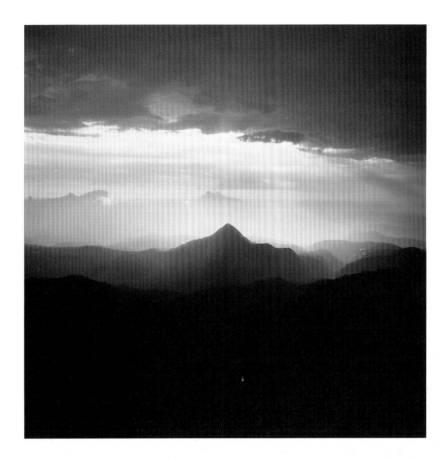

*Fig. 3.2.* Distant rain appears pale yellow as it scatters direct sunlight from overhead while the glowering clouds at the edge of the storm are a cooler, more neutral shade.

During the middle hours of the day, the Sun illuminates the atmosphere from horizon to horizon, creating the familiar blue embrace of the sky. But the blue light that gives the sky its luminosity has been separated from the sunlight that reaches the ground, so we see the Sun as slightly yellower (white light minus blue is yellow) than we would if we were above the Earth's atmosphere. Although we can see the two components if we look at Sun and sky together, it is much more convenient and revealing to look at their individual effects in a photograph.

A photograph also has the advantage that it removes the scene from its natural surroundings, which can easily play tricks with the astonishingly fickle sense of colour that human beings have developed. It is even more convenient if the picture contains pure white objects arranged to reflect the Sun and sky separately.

Among the best natural diffuse reflectors of light are the myriad tiny crystals in a fresh fall of snow. When sunlit snow is photographed under a cloudless sky its pure white surface recombines the direct rays of slightly yellow sunlight with diffuse blue light from the sky. The white that we see we register as 'cold', or at least neutral, because it is bluer than direct sunlight alone without the admixture of the blue from the sky.

The surface of the snow in effect recombines the scattered and direct components of sunlight to reflect the Sun's colour much as it would appear to an observer above the Earth's atmosphere. In the shade (beneath the trees in Fig. 3.1) we can see just how blue the snow appears when it is not directly illuminated by the Sun, though the shadows are under-exposed in the photograph and the colour is deeper than it would look to the eye.

A similar colour effect is seen beneath a cloudy sky, which is another natural, almost perfect diffuser. This time the diffuser combines the yellowed sunlight and blue sky from above by forward scattering rather than by reflection as with snow. The 'cooler' colour that emerges from the underside of clouds in Fig. 3.2 is thus nearer the true colour of the Sun above the Earth's atmosphere than is the yellowish hue of sunlight scattered by raindrops. Critical colour photographers make an allowance for the 'colder' light by using 'warming' filters on overcast days. Normally, we never register the fact, beyond applying the subjective word 'cool' to the feeling given by the slightly bluer light that filters to the ground through a canopy of clouds.

## THE SUNSET

As the Sun glides towards the western horizon at the end of the day, its colour and brightness are changed by the increasingly greater thickness of air that its rays have traversed. Because more of the shorter wavelengths are scattered away from the line of sight, the Sun appears ever yellower and fainter in Fig. 3.3, eventually becoming a rich red before it drops below the horizon. But the departure of the Sun does not immediately plunge the world into darkness, as it would if the Earth were airless. As it slips away, there begins one of nature's beautiful scenic displays, the changing colours of the sky during and after sunset.

One of life's more satisfying experiences is to wander around the walk-way encircling the dome of the AAT, watching the gathering twilight before a night's observing. From this vantage point, at Siding Spring Observatory, high above the western plains of New South Wales, the horizon is almost 100 kilometres away. A panorama of almost 360° can be seen, interrupted only by the nearby volcanic remnants that form the ancient, isolated peaks of Warrumbungle Range. One's joy is diluted if there is cloud around of course, but there is the consolation that it can sometimes provide an even more spectacular end to the day. But because they are no longer driven directly by solar power, clouds at night behave differently from those seen during the day, and the fluffy altocumulus pictured in Fig. 3.4 would almost certainly vanish as darkness falls.

Even as the Sun disappears in the west, the delicate reddish-purple of the Earth's shadow rises over the eastern horizon, its diffuse border merging into an ever-darkening blue-grey band. Overhead, the sky moves imperceptibly through deeper shades of blue until it merges with the Earth's shadow and the first stars appear. Beneath the walk-way, among the subdued colours of the Australian bush, kangaroos emerge to begin their nocturnal breakfast. Often, the last stages of twilight and the beautiful colours imperfectly captured in Fig. 3.5 must be foregone as duty calls and the final preparations for a night's observing are made. The twilight scattering of light from the nearest star by the components of the Earth's atmosphere provides a fitting introduction to the more distant but often equally beautiful parts of the Universe that will be examined later. Many of the same phenomena that are responsible for the colour of the Sun and sky are reproduced in distant clouds of gas and dust with equally colourful effect. The walk-way of the AAT is an ideal place to contemplate these links.

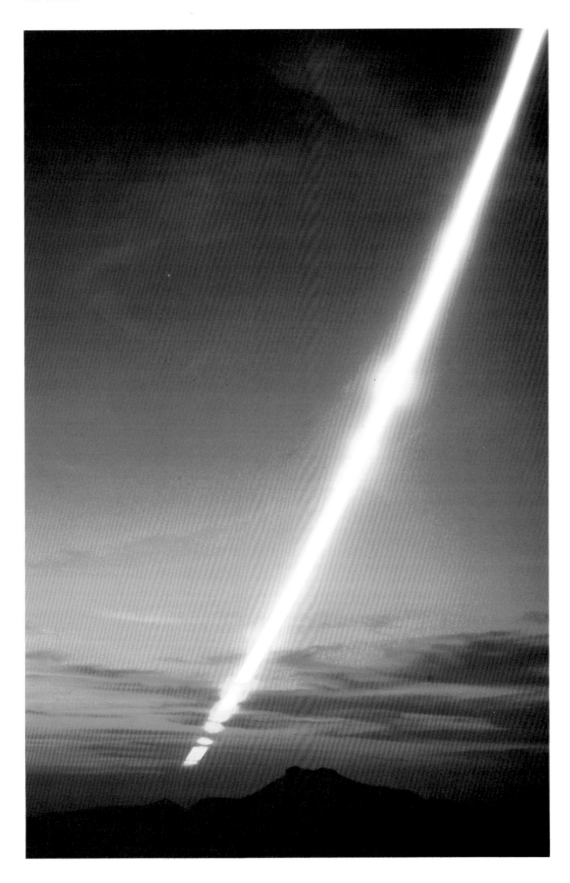

*Fig. 3.3.* The curved track of the setting Sun changes colour and dims as its light passes through an increasing thickness of air. This picture (and Fig. 3.2) was taken from the walk-way of the AAT dome and the distant mountains are those of the Warrumbungle National Park in New South Wales.

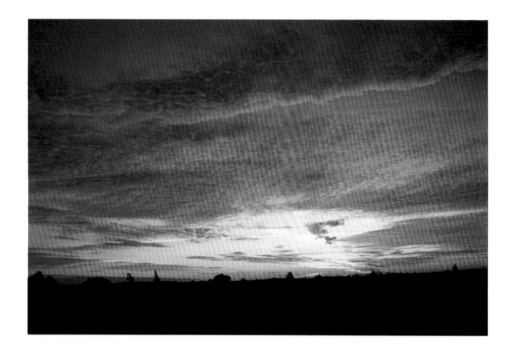

*Fig. 3.4.* Evening clouds make perfect reflectors for the colours of the setting Sun, which is well below the horizon from the camera position. Such clouds usually vanish after sunset, posing few problems for the optical astronomer.

*Fig. 3.5.* Normally, about an hour after sunset stars can be photographed in the zenith sky with an exposure of a few seconds. This picture was taken in 1987 and shows the colour of the horizon merging rapidly into near-darkness overhead.

Depending on the project, serious observing on a large telescope can often begin well before the end of astronomical twilight, which is conventionally taken to be when the Sun is 18° (about 36 Sun diameters) below the horizon. Since the Sun crosses the sky at 15° an hour, at the 31° S latitude of Siding Spring true astronomical darkness does not arrive until almost an hour and a half after the Sun has set, depending on the season, and recently, the high altitude particulate material from Mt Pinatubo has produced some spectacular sunset colours, which are seen in Fig. 3.6. With the vivid colours have come twilights that are unusually long, much to the consternation of astronomers who expected a dark night. But to most of us the sky appears dark long before astronomical twilight, and other kinds of darkness are recognised by the definitions of civil twilight, which ends when the Sun has sunk 6° below the horizon, and nautical twilight, which ends when it is 12° below.

Though there are many occasions when time permits the end of day and the beginning of night to be enjoyed in their entirety, these events also occur in reverse, though fewer people seem to notice, probably because dawn is a less convivial time than sunset. For thousands of generations, the onset of darkness has encouraged social activity as

*Fig. 3.6.* The spate of volcanic activity in the early 1990s spewed vast amounts of tiny particles into the upper atmosphere. These particles act as reflectors and are responsible for the vivid colour of many twilight skies of recent years and the extended duration of twilight. This photograph was taken in 1991 at about the same time after sunset as that in Fig. 3.5 and shows the twilight colours extending high into the sky (see also Fig. 3.17).

human beings have gathered for mutual comfort and collective safety, but at dawn the wise and the non-astronomers are still in bed.

To those used to looking at the night sky, the first glimmerings of the new day can be seen before the beginning of astronomical twilight. This is because, unlike sunset, the events unfold before eyes that are thoroughly dark-adapted. After what seems the darkest part of the night, the first brightening of the new day appears over the whole sky rather than being confined to the eastern horizon, and its perception is another example of the astonishing sensitivity of the human eye.

Inside the dome, perched at the top of the telescope is the prime focus cage of the AAT where I have spent many hours. From there, only a small patch of sky can be seen, framed by the $5 \times 6$ metre (about $16 \times 19$ foot) opening. Sometimes the pre-dawn brightening seems to occur surprisingly early. Then we discover that the planet Venus has risen, often a brilliant sight in the still-dark sky. From outside the dome it can be bright enough to cast shadows on the white paint-work. From within, its light is seen indirectly, scattered by the patch of sky visible through the dome opening immediately overhead.

## PHOTOGRAPHING THE SKY AT NIGHT

Most people can see well enough to move around freely at night by the light of the Moon alone. It is only when some visually demanding task is required that we realise that moonlight is almost a million times fainter than sunlight. Not only does the eye lose its sense of colour under these conditions, but the ability to perceive fine detail is diminished as well. However, the response to motion, especially that at the edge of the visual field seems largely unimpaired. This remarkable adaptation of the mechanism of sight to low light levels no doubt has survival value. Much better to see the tiger move than to recognise its colour or admire its stripes.

Just how remarkable an organ the eye is can also be demonstrated photographically. It is only when we try a snapshot of the night sky that we realise how faint the stars really are, which is one reason why we do not see them during the day. If the same photographic experiment is attempted when the Moon is full and high in the sky, we find that we need an exposure at least a million times longer than that needed during the day to capture an image of the landscape. This translates to exposures of 15 or 20 minutes with normal colour films and lens apertures, but the eye grasps the essence of the scene in an instant.

The photographic experiment in moonlight is worthwhile, however. It reveals that the moonlit sky is as blue as it is in the day, which is something the eye does not register. While the blue night-time sky may at first seem surprising, it is not unreasonable. After all, the Moon shines by reflected sunlight, and the molecules of the air are still there at night, just as capable of scattering the blue component of sunlight, despite the fact that there is much less of it and it has arrived by a roundabout route. But what is surprising is that the eye sees no hint of blue in the moonlit sky. Certainly we see the sky as much brighter when the Moon is near full, but there is no suggestion of colour.

Simple experiments with coloured card or cloth examined by the light of the Moon show that the eye is a very poor discriminator of hue under these conditions. Some pastel

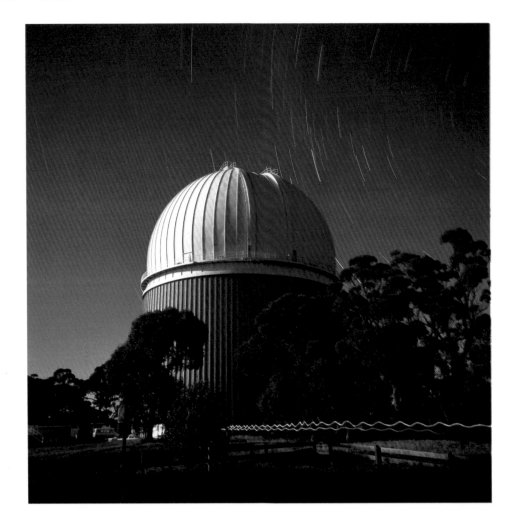

*Fig. 3.7.* The moonlit sky is every bit as blue as that seen by day, and for the same reasons, although the Moon itself is slightly yellower than the Sun. Its direct light is reflected on the dome of the AAT. Despite the moonlight, passing astronomers found it necessary to use a torch, leaving a wavy line as they walked past the camera.

shades can be identified but saturated colours and the moonlit landscape itself are seen as shades of grey. Nor is there any visual indication that the Moon is yellower than the Sun. But it is, and photographs of white objects such as the dome of the AAT pictured in Fig. 3.7 show this very well.

It is popularly believed that the Moon has an effect on human behaviour. While this seems to be another of those suburban myths, there is no doubt that it affects the habits of optical astronomers. On many large telescopes their schedules are divided up into 'bright' and 'dark' periods which are governed by the proportion of the night that the Moon is visible. Work on faint objects or that involving the blue end of the spectrum can only be done in dark time. Thus one is more likely to find astronomers whose research involves the infrared part of the spectrum working on nights of the full Moon, which only confirms what we always believed about infrared astronomers.

When the Moon is absent, the night sky is never completely dark, even on a mountain top far away from city lights. This can easily be demonstrated by holding up an outstretched hand to the sky. Even at the best astronomical sites, those selected for their darkness and clarity, its silhouette is clearly seen and the fingers are easily counted, and not merely because they hide the stars. Though the dark-adapted human eye can detect the uniform glow of the night with ease, it is normally so weak that its light is extremely difficult to photograph with ordinary photographic materials.

The natural luminosity of the night is a mixture of feeble radiation from many sources, some nearby, some extremely distant. Some of this light comes from unresolved stars in our own Galaxy and from the light of innumerable other galaxies far beyond. Some arises from sunlight scattered by small particles within the solar system but well beyond the influence of the Earth. A special case of this is the zodiacal light, discussed below. A small contribution is from the Rayleigh scattering of starlight by the atmosphere. But almost half of the night sky brightness, sometimes more, arises in the atmosphere itself, and is termed the air-glow.

The upper layers of the atmosphere are continuously bombarded by energetic outpourings from the turbulent surface of the Sun. Apart from visible radiation, some of this energy is in the form of ultraviolet light, which causes the air to fluoresce like a giant advertising sign. All radiation from the Sun arrives here a little more than eight minutes after it has left its source, i.e. during the day, so any visible effect this might have is lost in the Rayleigh scattering that dominates the daytime sky.

Other energetic arrivals from the Sun are the electrically charged particles that are part of its tenuous but extremely hot atmosphere that extends into the solar system, far beyond the orbit of the Earth. These particles can take days to reach us, and a steady stream of protons and electrons enters the upper layers of the Earth's atmosphere day and night, causing it to glow weakly.

The weak, natural air-glow is the price we pay for a warming Sun and protective atmosphere, but it complicates life for astronomers. It also places a limit on the duration of photographic exposures by eventually fogging the film or plate. At times when the Sun is especially active, sporting spots or flares, more particles are produced and the night sky is visibly brighter, a condition that can last for a substantial fraction of the 11-year cycle of solar activity. The night sky brightening caused by the solar activity can be so significant that it affects the astronomical recording of the faintest objects for years at a time and the recent 1989–91 maximum has been among the brightest for years. Not unconnected to the solar activity are the aurorae, from a photographer's point of view by far the most devastating of the night sky air-glow phenomena.

These too are the result of solar activity and are a fairly common sight at high latitudes, where the incoming particles are funnelled by the Earth's magnetic field into a narrow region around its magnetic poles. They are only rarely seen within the band of latitude 30° either side of the equator. However, several times during the recent spell of solar activity spectacular effects were seen as far north as Siding Spring on several occasions. The vivid red pictured in Fig. 3.8 lasted for an hour or more and reappeared over the southern half of the sky on many nights, a few days after massive flares had been detected on the Sun's surface. The red colour is from the excitation of oxygen and nitrogen in the upper atmosphere by energetic particles which, travelling at 1000 kilometres per second, had taken several days to cover the 150 million kilometres that separate us from the Sun.

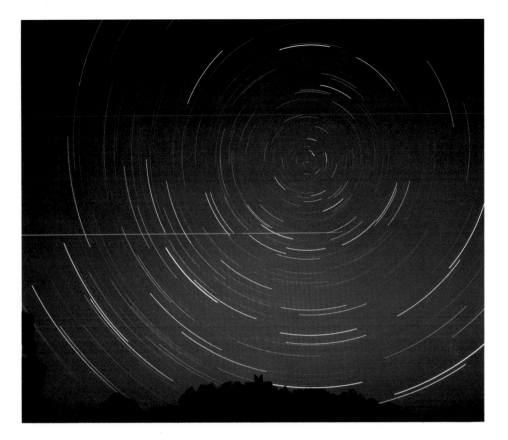

*Fig. 3.8.* Aurorae are rarely seen from the latitude of Siding Spring, 30° from the equator, but the intense solar activity in the 1990–91 cycle resulted in several displays as striking as this. The 30 minute exposure blurred out details in the flickering, curtain-like structures that were seen.

Only in a really dark sky does another solar system phenomenon, the zodiacal light (Fig. 3.9), become visible. It is most clearly seen in the west two or three hours after sunset, or in the east a similar period before the Sun appears, and only then if the Milky Way is in another part of the sky. Unlike the aurorae, the zodiacal light has nothing to do with the Earth's atmosphere. It originates in light scattered from a thinly populated band of relatively large particles orbiting the Sun roughly in the plane of the planets. The planets and the Sun itself appear to move through the zodiacal constellations, hence the name. The cloud of particles is regularly replenished by comets such as Halley's (Fig. 3.10), which leave a trail of solid debris behind as their volatile components are evaporated by passage through the inner solar system.

The keen-sighted can trace the zodiacal light high into the sky where it appears to brighten slightly into the 'gegenschein' (literally 'opposite light'). The gegenschein marks the anti-solar point, that place in the night sky where we look into the Earth's shadow in a direction exactly opposite to that of the Sun. However, unlike the sunset phenomena, this is not an atmospheric effect, but is sunlight reflected from dust in the solar system beyond the Earth's orbit. It is the celestial equivalent of the 'glory' or bright halo around one's own shadow (or more usually that of one's aircraft) seen as one looks down on sunlit clouds. The night-time version requires that we have good night vision, inhabit a very dark place and are well dark-adapted. Only then, and with some difficulty, might we see the anti-solar point as a feeble, diffuse glow in the sky.

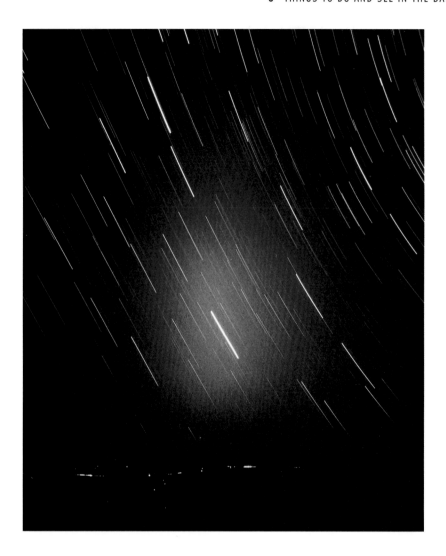

*Fig. 3.9.* The zodiacal light is sunlight reflected from small particles that are in orbit around the Sun in the plane of the planets. It is quite easy to see in clear, dark skies several hours after sunset, but its enormous extent and faintness makes it difficult to photograph.

The zodiacal light and gegenschein are only noticed in places that are naturally very dark. The vast, sparsely populated continent of Australia is an ideal location. Its 17 million inhabitants are mainly gathered around the south-eastern seaboard of an island that has an area about the same as that of the contiguous United States. Thus the interior is, by many standards, almost devoid of people, but even more importantly, it is free from the light and air pollution that gatherings of people seem to generate. The atmosphere is in any case naturally much cleaner in the southern hemisphere because of the relatively low population, small amount of industry and large expanses of cleansing ocean.

Siding Spring Observatory is in north-western New South Wales, Australia, about 355 kilometres north-west of Sydney, Australia's largest city. The darkness of the night sky is a valuable attribute for an observatory and once destroyed is unlikely to be recovered. A few years ago I made a night-time photographic survey of the whole horizon as seen from the mountain-top, using highly sensitive astronomical emulsions to record the faintest light. The experiment will be repeated in a few years to see if our precious darkness is threatened and to pinpoint where the threat might come from.

Sections from the resulting panorama, made over a couple of nights in 1986, are
duced as Fig. 3.11(a) and (b). The individual prints show numerous sources, many
ich are small towns far enough away to be below the horizon as seen from Siding
Observatory. The picture records the light from distant street lamps that is use-
directed upwards and then scattered towards us by the night air and tiny particles
ded in it.

lost of these glimmerings on the horizon are invisible to the naked eye but long
res and sensitive photographic materials help to reveal what the eye cannot. That
e detected at all is remarkable proof that Siding Spring is a truly dark observatory
he brightest artificial glow in its night sky comes from the nearest town,
arabran, which has a population of around 3500. It is 21 kilometres away and
ough for its lights to be seen directly. Fortunately, the local authorities are aware
otential problem that the town's lighting presents to the observatory, which is a
mployer and tourist attraction, and the township now has lighting codes on new
nents.

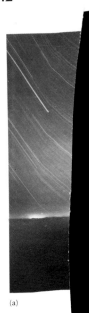

(a)

Rathe

and coastal

metres sout

surprising th

Special

tory sites. Th

excellent plac

2400 metres,

spilling into t

peaks on most

thin clouds of                                                                          rticu

when I visited the island to use the Isaac Newton Telescope in September 1991.

The dust is mainly composed of particles that are large compared with the w
length of light and does little to change the colour of light passing through it. It there
has only a small effect on most astronomical observations; indeed, it often seems
accompanied by warm, dry air that gives excellent stability to stellar images, like
cirrus cloud. Sometimes, however, when the lower slopes are free from cloud,
lighting is reflected by the dust and is detectable in the most sensitive observa
including photography. The scatter of yellow sodium light high into the sky is sho
Fig. 3.12, looking to the north-west, and the greener colour of mercury vapour in fl
cent lighting is seen to the south in Fig. 3.13. These pictures were made from arou
Isaac Newton Telescope.

## STAR TRAILS

Features common to both the previous photographs are the star trails, vivi
of the Earth's rotation. Star trails are amongst the simplest night-time pictures t
but are wonderfully revealing and beautiful. With an ordinary camera and col

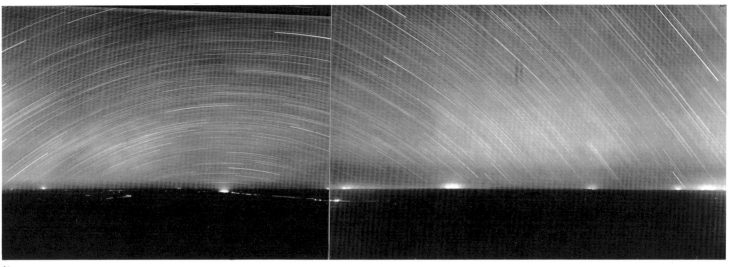

(b)

(b) None of the light sources revealed in (a) yet presents a threat to the telescopes at Siding Spring Observatory, and in all other directions hardly anything can be seen other than distant towns, most of which are invisible to the unaided eye. A few cars leave evidence of their movement across the landscape below. Here we are looking due north from Siding Spring.

one can produce stunning images that show the striking lines of stars setting, or the curved tracks of those stars that never set, gracefully wheeling around the celestial poles. It helps, of course, if there is a distinctive foreground to manoeuvre into view and if the night is long and dark.

These images are not merely pretty pictures. They reveal the underlying basis of celestial navigation, explain the gradual rotation of the seasons, and define the local latitude precisely. The stars also show themselves to be coloured, with a range of hues from sky blue to deepest orange. The astronomical implications of the colours are discussed in Chapter 4.

The photograph of stars rising over Coonabarabran shown in Fig. 3.14 was taken from the walk-way of the UK Schmidt Telescope on Siding Spring mountain. In the foreground are the lights of the town, 21 kilometres due east of the Observatory. The exposure was started when the distinctive line of three stars that form Orion's Belt was a few

*Fig. 3.12.* Saharan dust hides the horizon and scatters the yellow glow of sodium street lighting above Punta Llana, a small town at sea level on the island of La Palma.

*Fig. 3.13.* To the south of the La Palma observatories, the dust reflects the mainly fluorescent street lighting from Los Llanos.

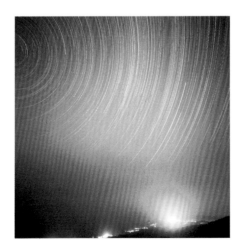

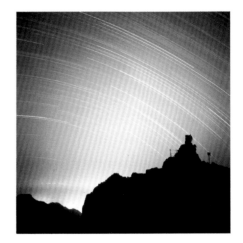

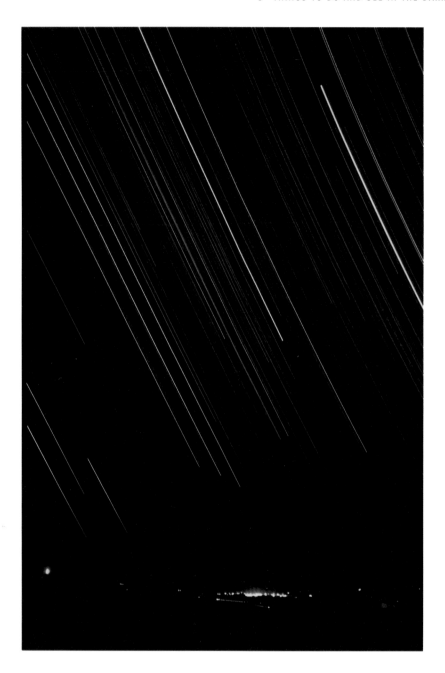

*Fig. 3.14.* The appearance of Orion in the morning sky heralds the imminent arrival of the southern summer, and its distinctive pattern of stars is seen here rising over the lights of the small town of Coonabarabran, which is due east of Siding Spring mountain. The colour film used for this exposure does not record the faint air-glow seen in Fig. 3.11(a).

degrees above the horizon. At the extreme left, to the north of the Belt, is the yellow trail of Betelgeuse, which can be followed almost to the horizon. To the right (south) of the three Belt stars is a fainter group of three, Orion's Sword. The central 'star' is the Orion nebula, which appears as a deep red streak. The bright trail that begins nearby and higher in the sky is that of Rigel which rose before the Belt. Just south of that is Saiph, the most southerly of the four bright stars that enclose much of this distinctive constellation.

The constellation of Orion must have been one of the first patterns of stars to have been recognised, and its annual reappearance in northern hemisphere skies is the well-

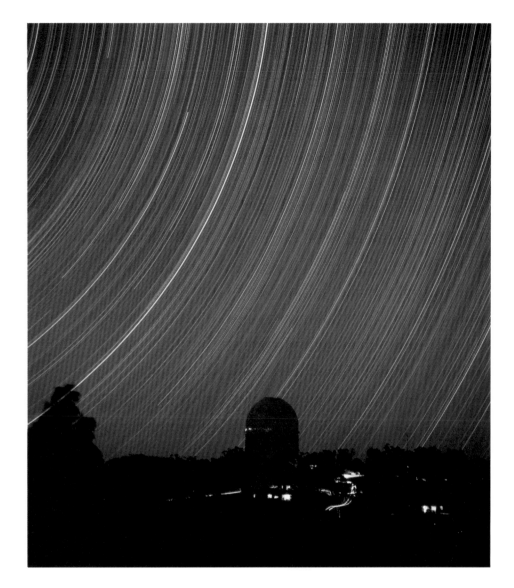

*Fig. 3.15.* The trails of stars seem to rain down as the Earth's rotation moves the camera beneath them. On the horizon, south-west of the camera, the rounded shape of the AAT dome is silhouetted against the night sky.

known sign that winter is near. To Australian astronomers, the return of Orion marks the end of the long winter nights when the brightest part of the Milky Way is overhead, and allows them to shift their attention to the Magellanic Clouds, where they can study the nearest galaxies. Unlike the far southern Magellanic Clouds, the stars of Orion straddle the celestial equator and can therefore be seen from all the inhabited parts of the Earth's surface. Because they are equatorial stars, their trails record as straight lines from any vantage point; however, the angle that the trails make with the local vertical changes with the latitude. In this case we can measure an angle of about 30°, which is the geographical latitude of Siding Spring.

The most prominent star image to appear in Fig. 3.14 is that of Sirius, the brightest star in the sky apart from the Sun. Its trail runs across the top right corner of the picture and is distinctly curved. Sirius is about 16° south of the celestial equator, and the more distant a star is from the celestial equator, the more curved its trail appears to be.

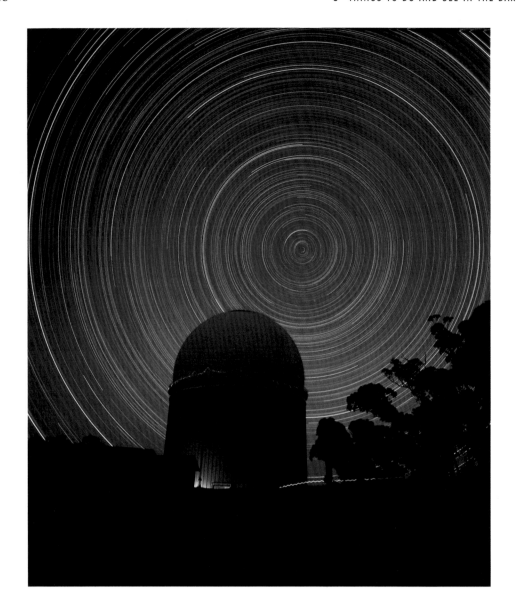

*Fig. 3.16.* Star trails wheel around the south celestial pole behind the AAT dome. This 9 hour exposure was long enough for the top of the building to be recorded by the light of the stars and night sky alone. Almost all the stars in this picture are circumpolar and so never set.

In Fig. 3.15 the same vantage point was used to make the picture, but the camera was turned to the south-west to record stars setting behind the dome of the AAT. Though I selected the scene and opened the camera shutter, it was closed just before twilight began by a colleague observing at the UK Schmidt Telescope. It is very handy to have friends whose job keeps them up all night! Because the exposure was much longer than that of Fig. 3.14, many more stars are recorded, together with the night sky airglow. The crowded field and long exposure makes it difficult to identify individual constellations, but it is evident from the increasingly curved trails that the camera was pointing further south than it was when Fig. 3.14 was exposed. Towards the left (south) of the picture, it becomes apparent that some of the stars will circle the south celestial pole without dipping below the horizon. These are the stars that never set.

In the view directly south (Figs. 3.16 and 3.17) the reason for this becomes clear.

*Fig. 3.17.* Careful placement of the camera produces star trails centred on the AAT dome. This exposure was started well after the end of astronomical twilight, but the volcanic dust still scatters some red light from the western horizon and affects the colour of the night sky.

The point around which the stars seem to revolve is (in our example) the south celestial pole, 90° around the horizon from the stars of Orion on the celestial equator. It is the place in the sky where we look out along the Earth's axis of rotation. If the picture had been taken in Antarctica, at the South Pole, the stars would seem to pivot around a point directly overhead, while most of Orion would circle just above the horizon.

Repeating this simple photographic experiment under more comfortable conditions at the equator would show the south celestial pole to lie on the southern horizon and the stars of Orion would produce trails rising and setting vertically from the eastern and western horizons. There would be a similar point surrounded by arcs on the northern horizon, exactly 180° away. There would also be a bright star, Polaris, very near the north celestial pole. It is seen as the smallest bright arc on Fig. 3.18(a), which was taken at the La Palma Observatory, about 30° north of the equator. The low contrast was caused by the Saharan dust, mentioned earlier, and I closed the shutter myself, being the scheduled observer that night. Fig. 3.18(b) shows the south celestial pole to be almost indistinguishable from that in the north, apart from the distinctive, stubby trail of Polaris.

The angular distance of the celestial poles above the true horizon therefore defines

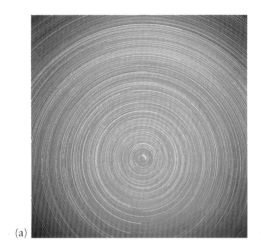

(a)

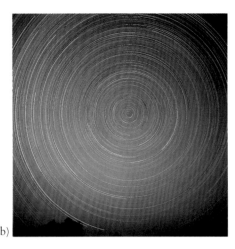

(b)

*Fig. 3.18.* (a) The north celestial pole is marked by a bright star, Polaris, the shortest arc, about 1° from the pole itself. The milkiness of the La Palma sky is mainly caused by airborne dust from the Saharan desert, together with volcanic dust at higher altitude. (b) shows the south celestial pole, photographed from Siding Spring using the same camera, lens, film and exposure time as (a).

the local latitude 0° at the equator, 90° at the poles and about 30° at Siding Spring, confirming the result from Fig. 3.14. If we point the camera due north from Siding Spring, we once again see the star trails as arcs, circling a pole that is now below the horizon. Just how far north we can see from this part of Australia is easy enough to calculate, and it turns out to be about +60° (−30° (overhead) + 90° (horizon) = +60°). We can see in Fig. 3.19 that the calculation is about right. The two bright stars trailing across the centre of the photograph nearest the horizon are Merak and Phecda, the northernmost stars of the Big Dipper, which are at about 55° from the celestial equator. The exposure time for Fig. 3.19 was about 4 hours, though because all the trails either start or end off the picture this information could not be obtained by measurement of the trails themselves.

In the case of Fig. 3.16, the exposure was much longer, about 9.5 hours. Since the angular extent of the star trails is visible, it would seem a simple enough exercise to calculate the exact exposure time by measuring the angle traced out by a single arc. If it were 90° the exposure time would be exactly 6 hours – after all, we assume that the Earth turns through 360° in a day. This simple assumption is not completely correct.

While it is true that the Sun seems to take exactly 24 hours to move through 360°, thus defining the day, it is not so for the stars. They appear to move through a full circle on the sky in 3 minutes 56 seconds *less* than 24 hours, because the stars keep sidereal time. This small difference exists because we are in orbit around the Sun and the length of the day is conveniently tied to that. However, this means that the Earth makes one more rotation in a year relative to the stars than it does to the Sun, so a solar day is 1/365th of a year (about 4 minutes) longer than a sidereal day. As a result, if we could see the stars during the day, the Sun would appear to move slowly against the starry background as the Earth moves around it, taking a year to return to the same place in the stars. Thus the pattern of the stars appears to change with the seasons as any given constellation rises almost 4 minutes earlier each night, and the star trails record sidereal time, the time used by astronomers to predict the positions of the stars night by night.

The long exposure used for the star trail photograph in Fig. 3.16 was enough to

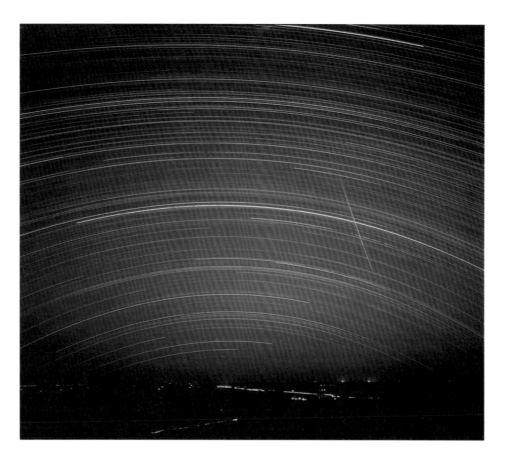

*Fig. 3.19.* The northern horizon as seen from Siding Spring is crossed by the trail of a bright meteor entering the Earth's atmosphere. The two bright stars near the horizon are in the Big Dipper.

record the colour of the natural diffuse glow of the night sky, which we said earlier was due in part to the excitation of the upper atmosphere by energetic particles from the Sun. Our picture shows it to be greenish, which is probably the result of strong, extremely narrow bands of radiation from excited oxygen. Other lines are also present, especially those in the red part of the spectrum from the hydroxyl radical (from water vapour) and from nitrogen. The relative strength of the red lines changes with different degrees of solar excitation, so sometimes the sky is slightly more yellow than green, as in Fig. 3.19. Sometimes, too, volcanic or wind-blown desert dust affects the appearance of the night sky, but only long photographic exposures reveal it.

As with the snow in the mountain-top scene described earlier, the white dome of the AAT building can act as an optical diffuser combining the yellow-green light from the sky and starlight. It so happens that most of the bright naked-eye stars are blueish in colour, a topic that is covered in Chapter 4. Pale blue and yellow-green thus combine to produce an almost neutral shade on the dome. Fig. 3.16 is one of the few pictures ever taken with an ordinary camera and colour film to show terrestrial objects illuminated by the light of the stars and the sky alone.

Fig. 3.20 shows an even longer exposure of the south celestial pole, this time in black and white. It was made using the same plate camera that produced the round-the-horizon pictures seen in Fig. 3.11, and reveals more stars than the colour picture because

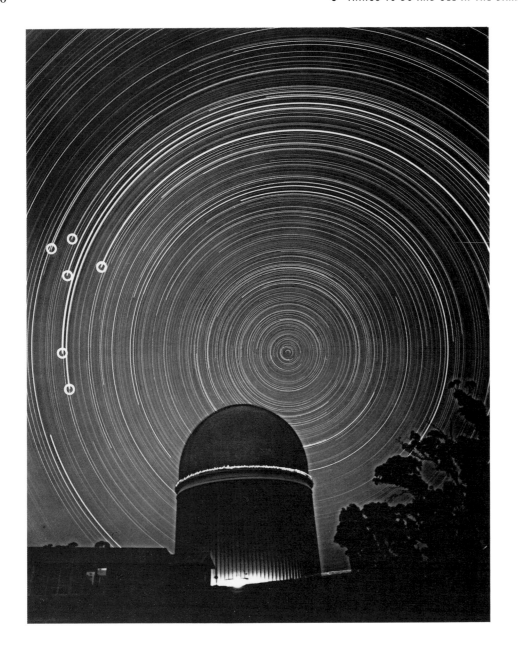

*Fig. 3.20.* The circles mark the position of the Southern Cross and Pointers at the beginning of this 10.5 hour exposure. The original plate has been copied in a special way to reveal the faintest stars, a process that produced the bright outline to the buildings and trees.

it was made using a special photographic emulsion designed to do just that. Unlike most of the pictures in the rest of this book, which have been made with the camera (or telescope) tracking the stars as they cross the sky, a longer exposure with a stationary camera does not mean that more stars are recorded. Instead, the star trails grow longer. Given Siding Spring's geographic latitude, and the necessity of starting and ending exposures when it is astronomically dark, the maximum length of night available in the winter is about 10.5 hours, which was the duration of the exposure for Fig. 3.20. There is nothing more unpleasant than climbing out of bed in the chilly, pre-dawn darkness of a winter's night simply to wander across a mountain top in search of a camera shutter that must be closed.

Such a hard-won picture does not record the brightness of stars equally over the whole photograph. Near the celestial pole, images are moving very slowly over the plate and so the effective exposure is longer and fainter stars appear. However, this part of the sky is something of a celestial desert and few bright stars are found there. In the northern hemisphere the brilliant star Polaris is only 1° away from the north celestial pole, which therefore requires no further identification for navigators. In the south there is no bright star close enough to the pole to provide a convenient beacon. Instead we have a whole constellation to guide us. Although it is small, Crux is so conspicuous that it forms part of the national emblems of several austral nations. And as if Crux itself was not easily recognised, it has two Pointers to direct the eye.

The Southern Cross and Alpha and Beta Centauri will be familiar to anyone who has looked at the southern sky. Because they are quite close to the pole they can be seen at some time during any night of the year, unlike the stars of Orion, which are seasonal. Alpha and Beta Centauri are identified on Fig. 3.20, which shows why together they have the familiar name of the 'Pointers' to the Southern Cross. The picture also reveals how the long axis of the Cross points towards the pole and how its angle to the horizon changes through the long August night.

The apparent motion of the stars across the sky tells us more about the Earth than about the heavens, and an understanding of this is essential to the art of terrestrial navigation. But the bright stars of Crux and the Pointers in Centaurus which decorate the southern Milky Way, and the distinctive grouping in Orion, are all part of a huge system of stars known as Gould's Belt. This in turn is a manifestation of the galaxy of stars in which we live. We will look at this aspect of night-time photography in the next chapter.

# 4 THE MILKY WAY AND THE COLOURS OF THE STARS

## DISCOVERING STAR COLOURS

The colours of the stars are subtle but full of meaning. They reveal the temperatures at the star's surface and sometimes the state of its atmosphere. The surface temperature hints at conditions deep inside and the colours can tell of the age, size and surface composition of a star, and help to determine its distance. A star's colour can also reveal its status within a family group. Unfortunately, the human eye is scarcely able to see these subtle shades; though five orange and yellow stars have been known since antiquity, the discovery that most of the bright naked-eye stars are pale blue had to await the serious application of photography to astronomy in the 1890s.

For most astronomical activity, photography had displaced the visual observer by the turn of this century. Until then, the human eye was the only way of estimating the relative brightness of stars. This could be done quite precisely by experienced observers, who expressed their results on a magnitude scale, where each magnitude step corresponded to a change of apparent brightness by a factor of 2.512. But when the first photographic astronomers compared the visual magnitude of a star with its magnitude measured from a photograph they were puzzled to find that some stars appeared brighter on the photograph than they did to the eye, and, strangely, this seemed particularly true of many of the most conspicuous stars.

The answer to this puzzle lay in the colour sensitivity of the photographic emulsions. The earliest plates could only respond to blue light and invisible ultraviolet wavelengths, whereas the eye works best in the yellow-green part of the spectrum. What the photographs were showing was that an unexpected number of stars recorded strongly on the photographic plates and were therefore predominantly blue, and even more surprising, the number of blue stars seemed greatest among the bright naked-eye stars, those that made up the outlines of the constellations.

After centuries of sky-watching with the unaided eye, no-one had ever described the stars as other than white, yellow or reddish-orange. Some blue stars were known from telescope observations, but they were always members of strongly coloured pairs of visual double stars and were therefore a curiosity. Now, as so often happens, the arrival of a new technology had opened a new window on the Universe, which allowed the stars to be seen, literally, in a new light. The fresh approach that photography provided

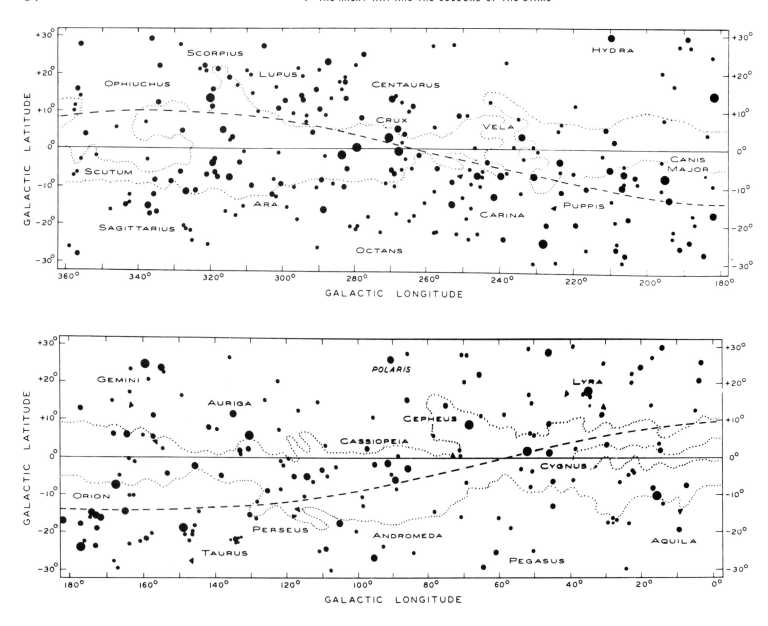

quickly led to new ideas about stars and galaxies and gave birth to the science of astronomical photometry, the measurement of the quality and quantity of starlight from their
recorded, as opposed to their perceived image.

As a result of this photographic discovery, it was soon remarked, mainly by the
American astronomer Benjamin A. Gould and his collaborators, that the predominantly
blue stars formed a broad belt somewhat inclined to the diffuse span of the Milky Way.
The centre-line of this rather straggling band of stars is the dashed line on Fig. 4.1.
Nowadays it is known as Gould's Belt, though it had been noted centuries before that the
very bright stars were concentrated towards the diffuse band of the Milky Way. The difference now was that photography had shown that a large number of the stars of
Gould's Belt were predominantly blue. However, the full significance of that apparently
simple observation had to await developments in other areas of physics.

*Fig. 4.1.* The brightest individual
stars of Gould's Belt are shown as
black blobs in this diagram, and
form a broad, irregular band whose
centre is marked by a dashed line
inclined at about 15° to the Galactic
equator, marked by the straight,
solid line. The extent of the Milky
Way's diffuse background is enclosed
in a dotted outline.

Now that the reason for the discrepancy between the photographic (i.e. Blue) and Visual (yellow-green) magnitudes was understood, it became clear that the *difference* in the magnitudes $(B - V)$ was a measure of the colour of a star and that this in turn was an indication of its surface temperature. The colour difference is known as the 'colour index' and can be expressed numerically. The standard $B - V$ colour index ranges from 2 or more for the coolest, reddest stars to −0.6 for the hottest, bluest. The Sun has a colour index of 0.62 on this scale.

The general mathematical relationship between the colour, temperature and surface area of a radiator of energy was quantified by Max Planck in 1900. Planck had to invent the notion of a 'black body' (and the concept of the quantum) to link these ideas, which are central to understanding the transfer of thermal radiation. His equation revealed that at very high temperatures incandescent bodies produced blueish light while cooler objects were yellow or reddish. This is of course contrary to the artist's understanding of colour, where blue represents a cool hue and red is associated with the warmth of the fire.

In astronomy, these concepts of colour have been refined into the elegant graphical comparison of colour and magnitude which was devised independently by Ejnar Hertzsprung and Henry Norris Russell in the years before the First World War. The Hertzsprung–Russell diagram is itself central to modern astrophysics and leads naturally to a system of stellar classification that links the colour and brightness of a star to its evolutionary state and to its mass.

From those relationships we now know that the sparkling jewels of Gould's Belt are extremely massive stars, which range in temperature from less than 3000 K (degrees kelvin) to 40 000 K or more. These are the giant and supergiant stars. Though their temperatures correspond to markedly different colours, with the hottest as blue as the daytime sky and the coolest a distinct orange-yellow, such stars are almost all quite young. Their ages are measured in millions of years, sometimes even less, compared to the thousands of millions of years of many less massive stars like the Sun. Though these brilliant youngsters seem to dominate the night-time sky, they are extremely rare. They appear numerous because they are many thousands of times more luminous than the much more common, intrinsically faint Sun-like stars and we can therefore see them over enormous distances.

## REVEALING THE COLOURS OF THE STARS

The star trail pictures in the previous chapter reveal the presence of many stars, but the colours of only very few. Another long-exposure picture of the sky, Fig. 4.2, was made by tracking the stars carefully for 45 minutes, not through any skill of mine, but simply by attaching a Hasselblad camera to the prime focus 'cage' fixed to the top of the AAT while I was taking long exposure plates of something in that direction. Because the AAT is a superb tracking mount, complete with auto-guider, star images remain in the same place on the photographic emulsion for the full duration of the exposure, rather than slowly trailing across it.

*Fig. 4.2.* A wide-angle view of the constellation of Orion. Even a short exposure with an ordinary camera and colour film shows many more stars than the eye can see, so many that the principal stars of this very distinctive constellation are hard to identify.

The picture therefore shows many more stars than the trailed picture but even fewer now seem to be coloured. This is because the tiny images of the stars are very bright and, being in one place on the film for a long time, over-expose the emulsion. As in everyday photography, over-exposure results in washed-out, colourless images. The bigger images are brighter stars, their imprint swollen only because more of their light is scattered within the photographic layer itself. Very few star images receive just the right exposure to capture their colour. These appear as the smallest stars on the photograph.

However, by a lucky mishap, I discovered that the size of the impression left by the stars on the photograph can be changed deliberately, and the colour of some of them recorded, by the rather unexpected ruse of photographing them out of focus. Working in the dark and at a difficult angle in the prime focus cage, I accidentally shifted the focus of the camera lens rather than the aperture ring and began a new exposure with the focus set to a few metres instead of infinity, which is where the stars are (optically speaking). When point-like images are out of focus, their energy is spread over a greater area of the

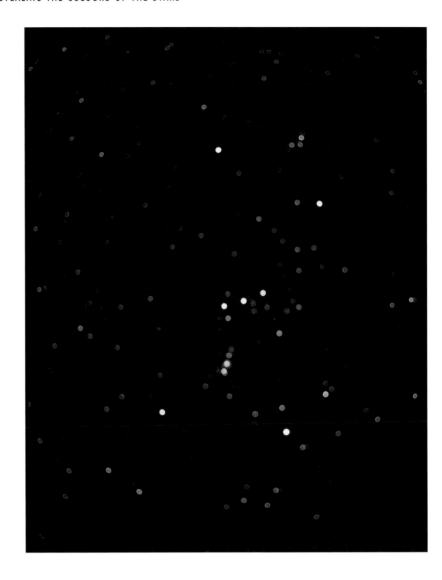

*Fig. 4.3.* The same part of the sky as Fig. 4.2, but with the camera lens focused on about 3 metres. The faintest stars have vanished, but the out-of-focus images show the colours of some of intermediate brightness. The bright stars that mark this well-known constellation are now clearly seen.

film, so the faintest stars recorded in Fig. 4.2 are now severely under-exposed and thus invisible, but in Fig. 4.3 some of those of intermediate brightness begin to reveal their true colours. Even here, the brightest are still burned-out and appear white. We now see that Figs. 4.2 and 4.3 are of the same part of the sky, the familiar grouping of stars that make up the constellation of Orion, first seen above Coonabarabran in the previous chapter.

This silly error led me to devise a surprisingly direct and novel way to photograph star colours that requires nothing but a simple camera and a firm support such as a tripod. The trick is to alter the focus of the lens in a series of steps *during* a star trail exposure. This is done while the camera is in a fixed position and it results in the striking effect shown in Fig. 4.4. This photograph is a 30 minute exposure and shows the colours of all the stars in Orion that are visible to the unaided eye, and many that are not. Further details of this simple but revealing step-focus photography are given in the references in the Bibliography.

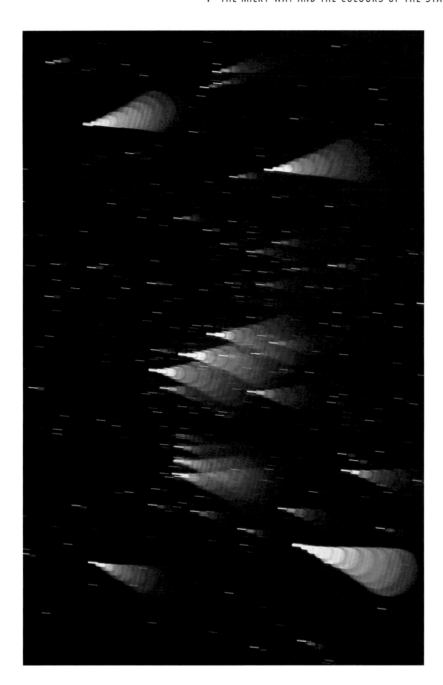

*Fig. 4.4.* The colours of all Orion stars down to about the eighth magnitude, well below naked-eye visibility, are revealed in this photograph. It was made by fixing the camera and allowing the stars to drift across the field of view while the camera lens was progressively de-focused. The red streak is the Orion nebula itself while Betelgeuse is the deep yellow star at upper left.

In Fig. 4.4 the photograph has been oriented with north at the top and east to the left, as is the normal astronomical convention. It shows stars from well below the threshold of naked-eye visibility to the striking yellow-orange of Betelgeuse and dazzling blue of Rigel, both seemingly a thousand times brighter. Of the several hundred stars visible in the photograph, all the brightest are clearly blue, with the obvious exception of Betelgeuse (top left) and the bright red streak of the Orion nebula. This is the central 'star' in the line of three forming Orion's Sword Handle, which hangs from the brighter and very distinctive line of three similar stars that mark his belt.

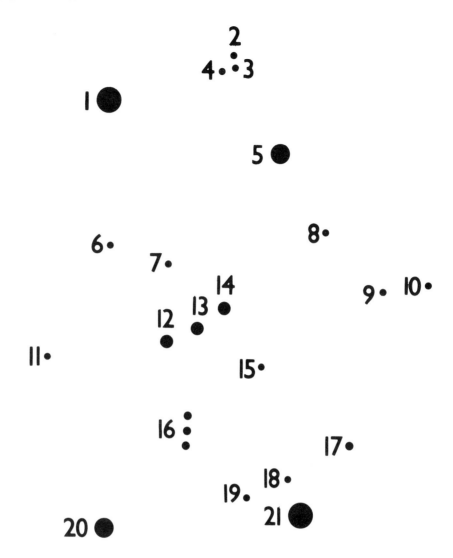

*Fig. 4.5.* The identification of the brighter Orion stars shown in Fig. 4.4. The numbers refer to the stars listed in Table 4.1.

The stars of Orion's Belt are Alnitak, Alnilam and the extremely hot Mintaka at its western (right) end. Both Mintaka and Bellatrix, in Orion's left shoulder (top right of the picture) are hotter than the other prominent stars in the picture and therefore appear to be a slightly deeper blue.

These star names, and most others in this easily recognised constellation, were recorded by Arabian astronomers many centuries ago, and are enduring reminders that astronomy (literally star-naming) is an ancient art as well as a modern science. More recently, that is since 1603, the order of the apparent brightness of stars in many constellations has been indicated by Greek letters. In some, it is an order of position; like the naming and delineation of the constellations this system too is far from systematic. The stars in Fig. 4.4 are identified in Fig. 4.5 and listed, together with their spectral types, colour indices and approximate distances in Table 4.1. The colour indices are in the conventional $B-V$ colour index system, where large numbers correspond to cooler (and thus yellower) stars and small (and especially negative numbers) refer to hot, blue stars. The

Table 4.1. *Bright stars in the Orion region*

| No.[a] | Name | Visual magnitude | Colour index | Spectral type | Distance (light years) |
|---|---|---|---|---|---|
| 1 | α Ori (Betelgeuse) | 0.50 | 1.85 | M2 Ia | 310 |
| 2 | λ Ori (Meissa) | 3.39 | −0.18 | O8 | — |
| 3 | $\phi^1$ Ori | 4.41 | −0.16 | BO IV | 1860 |
| 4 | $\phi^2$ Ori | 4.09 | 0.95 | KO III | 195 |
| 5 | γ Ori (Bellatrix) | 1.64 | −0.22 | B2 III | 360 |
| 6 | 56 Ori | 4.78 | 1.38 | K2 II | 780 |
| 7 | 51 Ori | 4.91 | 1.17 | K1 III | 285 |
| 8 | ρ Ori | 4.46 | 1.19 | K3 III | 280 |
| 9 | W Ori | 6.17 | 3.45 | C6 II | — |
| 10 | $\pi^6$ Ori | 4.47 | 1.40 | K2 II | 620 |
| 11 | — | 4.53 | 1.22 | K2 III | 260 |
| 12 | ζ Ori (Alnitak) | 1.77 | −0.21 | O9.5 Ib | 1110 |
| 13 | ε Ori | 1.70 | −0.19 | B0 Ia | 1210 |
| 14 | δ Ori (Mintaka) | 2.23 | −0.22 | O9 II | 2400 |
| 15 | η Ori | 3.36 | −0.17 | B1 V | 750 |
| 16 | Group of O and B stars and the Orion nebula | | | | |
| 17 | β Eri (Cursa) | 2.79 | 0.13 | A3 III | 91 |
| 18 | τ Ori | 3.60 | −0.11 | B5 III | 425 |
| 19 | 29 Ori | 4.14 | 0.96 | G8 III | 190 |
| 20 | κ Ori (Saiph) | 2.06 | −0.71 | B0.5 Ia | 68 |
| 21 | β Ori (Rigel) | 0.12 | −0.03 | B8 Ia | 915 |

[a]The stars are identified in Fig. 4.5.

spectral type is a shorthand way of identifying the basic properties of stars based on the position and strength of various spectral lines. The classes OBAFGKM are a series of decreasing surface temperatures from over 30 000 K (blue) through 6000 K (white) to 3000 K (yellow). To the initiated, the Roman numerals indicate six classes: Ia–c supergiants, II bright giants, III giant, IV subgiant, V main sequence and VI and VII white dwarfs. If the Sun were listed here it would be a G2 V star.

　　With modern knowledge of their spectral classification and distance, we can confidently say that the bright stars that we see in the photograph are young, massive and intrinsically very luminous. That a small fraction of the bright stars is yellow-orange rather than blue is also not unexpected, but not easily demonstrated without a simple photographic device such as this. These are young stars that have evolved to their red giant (or supergiant) stage, a relatively quiet interlude towards the end of the short but vigorous life of a massive star. Since the red giant phase is brief, and is even briefer for

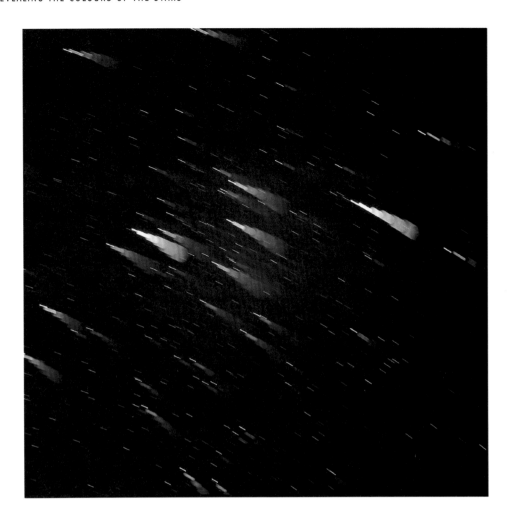

*Fig. 4.9.* The colours of the stars in the tail of the Scorpion. The brightest image here is that of the giant planet Jupiter (right) with Antares as the next brightest yellow star trail.

ing, abstract pattern is obtained if fewer steps are used, as in Fig. 4.7, but here the identification of the individual stars is more difficult. As before, a diagram (Fig. 4.8) helps, and the identified stars are listed in Table 4.2. In Fig. 4.9 we see the colours of the bright blue stars in the curved tail of the Scorpion that gives its name to the constellation, encircling the bright orange trail of Antares. The similarly bright, pale yellow interloper at the left of the picture is the planet Jupiter, loitering in the Milky Way when this picture was taken.

The nearest of all the bright stars (except of course the Sun) is Alpha Centauri. This multiple system is the most easterly (left) of the two Pointers. Apart from Alpha Aurigae (Capella), Alpha Centauri is the only bright star in the sky with a colour similar to that of the Sun. Alpha Centauri therefore appears colourless in the step-focus picture, irrespective of exposure level, giving me confidence that the surprising colours recorded by the daylight-balanced colour film are an accurate reflection of nature. The lack of colour in Alpha is in sharp contrast to the sky blue of Beta Centauri which, while it appears only half a magnitude (33%) fainter, is over 100 times more distant and therefore about 10 000 times more luminous.

This observation underlines the enormous difference in the intrinsic luminosities of massive, young, extremely hot stars such as Beta Centauri when compared with less massive stars such as the Sun and Alpha Centauri. This simple experiment reveals in a dramatic way something that is not apparent to the human eye and that was only dimly perceived by Gould and the pioneers of photographic photometry: the bright stars of Gould's Belt are relatively distant and mostly blue with a few bright orange-yellow stars, which belong to the same young population, scattered between.

The colour–magnitude diagrams of modern measurers of star colours show that highly luminous blue stars (for historical reasons identified as classes O and B) are extremely rare, and that the yellowish 'bright giants' and supergiants that they evolve into are even rarer. Such stars therefore have short but brilliant careers, which usually end in the most catastrophic failure. Massive stars destroy themselves as supernovae, a topic covered in Chapter 7. By comparison, intrinsically faint, low-mass stars similar to the Sun are vastly more numerous, mainly because they survive for billions, rather than millions of years. They end their lives in a less dramatic way, as we shall see. But it is the few brightest stars that dominate the naked-eye appearance of the night sky, and the multitudes of lesser stars are hardly seen, insignificant contributors to the visible pattern of the constellations.

## THE COLOURS OF GALAXIES

The stars that pepper the night sky are all members of a vast spiral system of stars within which we live. They belong to *the* Galaxy, described with unwarranted familiarity as 'our' Galaxy, or the Milky Way. In it there are hundreds of millions of stars. From our vantage point on this planet, even the most powerful telescopes locate nothing like that number. We can therefore explore only a small fraction of the Milky Way and see little to suggest that it has a spiral structure. The reasons for this will become apparent in Chapter 5. However, the Milky Way is a galaxy rich in the ingredients of star formation and it contains many young stars. The most massive and most luminous of these stars are blue.

The link between the colours of the bright stars we see around us and our location in the spiral arm of a predominantly blue galaxy was not made as soon as blue stars were discovered. For a start, at the turn of the century, the nature, distance and dimensions of the objects then described as spiral nebulae were not known, and it was not firmly established until the 1940s that there was a colour gradation from blue to yellow within these systems.

While the earliest discoveries concerning colour in galaxies were made by observing stars in the Milky Way, the overall distribution of colours in galaxies is better seen by looking at a spiral system beyond the Milky Way. In Fig. 4.10 we see M83, one of the finest and nearest examples of a spiral galaxy anywhere in the sky. Because by chance we see from above one of its poles, it is face-on, and ideally placed for us to study the anatomy of a stellar system that is similar to the Milky Way. In our colour picture, the abundance of light from the blue stars in the arms of a spiral galaxy becomes vividly apparent.

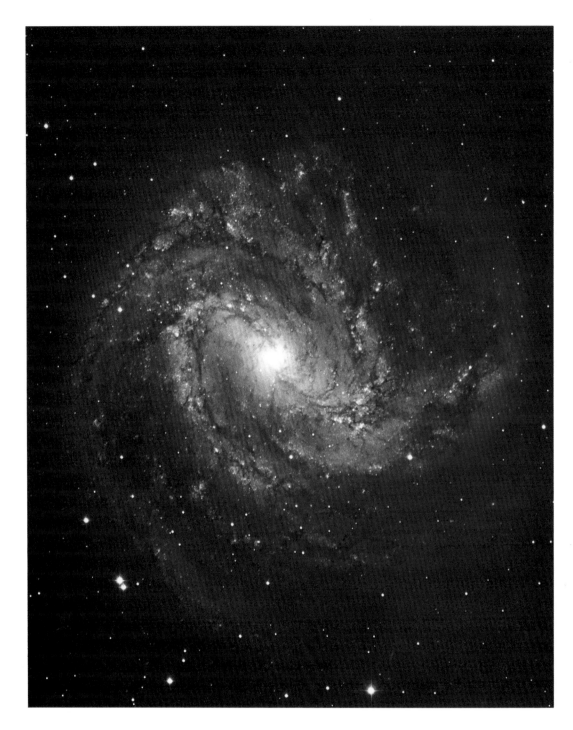

*Fig. 4.10.* The rich colours of M83 (NGC 5236) are revealed in this portrait of a galaxy believed to be rather like the Milky Way. M83 is on the border of Hydra and Centaurus at a distance of about 10 million light years and is one of the finest face-on spiral galaxies in the sky.

At a distance of about 25 million light years, M83 is close enough for a few of its most luminous stars to be seen as individuals. That we see only a handful of them in a galaxy of a thousand million stars is vivid testimony to their rarity and brilliance. As in the Milky Way, these young luminaries dominate the light output of the spiral arms. The spiral arms are also host to many recently-formed clusters of stars and they give the outer regions of M83 a clumpy appearance and a distinctly blueish hue. As in the Milky Way,

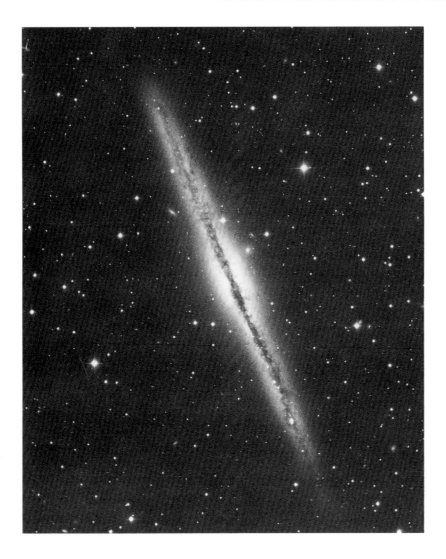

*Fig. 4.11.* If we could somehow view the Milky Way from some tens of millions of light years off the end of one of its arms we would see something like NGC 891, an edge-on spiral galaxy in Andromeda. The dark lane is dust, tightly bound to the plane of the galaxy.

young stars such as these often begin their lives in compact clusters, and the presence of bright knots of blue stars in the spiral arms emphasises the extreme youth of these distinctive features.

Within and between the arms are hosts of fainter stars of relatively low mass, and fragmented dust clouds, left behind after the curved wave of star formation passed through the system. The ranks of the faint stars will be augmented by many such cycles of star formation, but the short-lived, massive stars soon disappear as supernovae, which is why the brightest parts of the spiral arms are so relatively narrow, like the curved, advancing front of a bush fire leaving short-lived glowing embers behind. Many supernovae have been seen in M83, a consequence of its favourable orientation, its distance and its vigorous rate of star formation.

Forming the inner curves of the spiral arms are large numbers of pinkish nebulae, each one a region where new-born stars remain intimately associated with the gas and dust in which they formed. Most of this star-forming stuff is hydrogen, the most abundant element in the Universe. It absorbs ultraviolet radiation from nearby hot stars which is then re-radiated at a longer wavelength, as a red colour characteristic of the simplest of all the elements.

Finally, on the inside of the spiral arms and closely associated with the visible nebulae, there are found long straggling clouds of dust, seen as dark yellow-brown in this colour picture. The dust is evidence of previous episodes of star formation and destruction that have enriched the galaxy with the heavy elements that are capable of forming solid particles. Swept up into dense clouds, the dusty remnants of dead and dying stars are about to be compressed by the advancing spiral arm into new stars, and possibly planets too.

The hazy central region of M83 has a colour and texture quite unlike that of the spiral arms. Around the nucleus of the galaxy there are relatively few of the highly luminous stars that produce so much of the light in the outer arms. Instead, there is a high concentration of stars that are old, cool and yellow. Although these stars are individually much fainter than the sprinkling of blue giants further out, towards the centre they are much more numerous and their light dominates a large region around the quite small nucleus.

This colour picture confirms what we already know about the general distribution of the stars and other ingredients of spiral galaxies in a spectacularly satisfying way. It was technically possible to make such a colour picture in about 1910, but it was 1942 before clear proof of the colour differences seen here was laboriously extracted from two-colour (B and V) plates of our own Galaxy by Walter Baade. The three-colour image of M83 was made from plates I took with the AAT in 1985 and was among the first to show clear colour differentiation in a spiral galaxy.

Like the Milky Way, M83 is a flat disk. If we were able to turn it on its side, we would see just how slender it really is. We would also see that the dust lanes found in most spirals are tightly constrained to the galactic plane. Of course, we cannot turn M83 around to look at its profile, but spiral galaxies are numerous and are seen at all angles. Though some can be found that are suitably oriented, statistics ensure that only a very few are seen exactly at right angles to their rotational axis. The best edge-on galaxies are in the northern hemisphere, out of reach of the AAT, but a visit to the Isaac Newton Telescope on La Palma enabled me to take a series of plates of perhaps the finest example, NGC 891, Fig. 4.11.

The picture shows the colour effect produced by the dust lane. As we shall see later, this yellow-brown hue is typical of the absorption of starlight by very small particles. Astronomers call this 'reddening', and while it is true that the light emerging from the dust is richer in red than the light that entered it, no red light has been added. A more precise but less memorable name for this phenomenon would be 'de-blueing'.

The particles that are responsible for de-blueing light are collectively described as dust, but they are more like the smoke of the fire than the dust of the desert. Where they are most numerous no light passes at all and a dark line seems to divide the galaxy into two and hides the star-forming regions close to the remarkably narrow plane of the galaxy. Usually less than 1% of the mass of a spiral system is in the form of dust, yet it seems to be an important ingredient for creating stars, and is essential for making both planets and people. Unfortunately, it also hides from us most of our own Galaxy, the Milky Way.

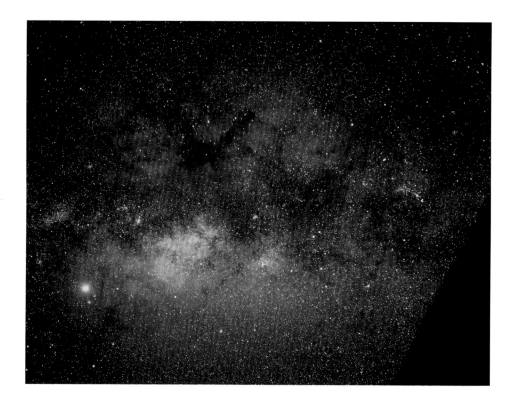

*Fig. 4.12.* A spiral galaxy seen from within. This wide-angle view of the Milky Way is the best we are likely to have at visible wavelengths. It was taken with an ordinary camera attached to the end of the AAT while the plates for Fig. 4.18 were being exposed.

## A LOOK AROUND IN THE MILKY WAY

How much is hidden by the dust of spiral galaxies will become evident if we take an imaginary journey towards the heart of the edge-on galaxy NGC 891 in the plane of its flat disk. As we move in towards the nucleus, we will pass through the patches of obscuration that line the spiral arms, and the rest of the Galaxy will occasionally disappear from view as we enter a chilly cosmic cloud. Though the tiny smoke particles that obstruct our view are very few and far between, the Galaxy is huge and in some directions the particles are numerous enough along the line of sight to hide the way ahead completely. Sometimes the dust is so thick that we can see for less than a light year around us and only one or two stars light the blackest of skies. As we move inwards from the darkness, we might re-emerge into a region dominated by nebulae that glow with the radiation from hot stars, occasional beacons among the vast, tenuous clouds of smoke and gas stirred up by their activity. We still cannot see very far, perhaps a few light years, but beyond this turbulence, as the number of very bright stars begins to thin, we enter the relatively clear, quieter region between the spiral arms.

Though we can now see for much greater distances on either side, and the view above and below the Galactic disk is hardly obscured at all, ahead are more billowing smoky clouds. Those that are nearby seem to be lifted high out of the narrow plane of the disk by the effects of perspective, like terrestrial clouds advancing from a distant horizon. Far ahead is yet more dust, part of an inner spiral arm, its irregular cumulus completely blocking our view towards the nucleus of the galaxy. We see all this against a faint backdrop of masses of distant stars and through a veil of nearby ones, very few of which are the spectacular kind. Like any traveller entranced by the view, it is natural to stop and take a picture. We must be patient if we want to record the faintest parts, and the exposure will be lengthy, but the result is worth while.

The photograph might look like that reproduced in Fig. 4.12. But this is not a snapshot of the inside of any random galaxy, or even NGC 891. This is the view towards the centre of the Milky Way, and it was photographed from Siding Spring, with a camera attached to the prime focus cage at the end of the AAT. The panorama covers about 45° of the southern sky and shows the only perspective we are ever likely to have of our Galactic neighbourhood. Here our imaginary journey must end, for we will never be able to travel onward to the nucleus of the Milky Way, which we cannot in any case see at all at optical wavelengths. The next spiral arm inwards, the so-called Sagittarius Arm of our Galaxy, hides it from view.

While the nucleus itself is beyond the range of optical astronomers, the cloud of mostly faint stars in orbit around it – the bulge – is only partly obscured, at least in that part that lies towards the southern constellation of Sagittarius. Much of the northern part of the Galactic bulge is almost completely hidden from us by the nearby dust cloud mentioned earlier. What little light does penetrate is very deficient in blue, 'reddened', not to be confused with the Doppler redshift due to movement of the light source away from us. Extreme reddening actually yields faint, brownish hues, and that is how we see the northern bulge, almost out of sight.

The resolved stars scattered uniformly across the picture are mostly members of the 'home' spiral arm or faint inter-arm stars of which the Sun is the nearest example. We look through this veil of nearby stars to a more distant vista in Sagittarius, which we surmise is also part of a spiral arm. Here and there are red emission nebulae; those on 'our' side of the Sagittarius Arm are the most prominent. On the left of the picture is M8, the Lagoon nebula, and M20, the beautiful Trifid nebula and their associated star-forming regions, which are seen in more detail on pp. 118–26. On the right of the picture, to the west of the obscured nucleus of the Galaxy, are two fainter emission nebulae, NGC 6334 and 6357, embedded deep in the dust, perhaps within or beyond this inner spiral arm. These are seen more clearly in Fig. 4.13. If the centre of the Galaxy were visible, it would appear near the middle of an imaginary line joining M8 and NGC 6334–57. The location of the objects identified in Fig. 4.12 is shown in Fig. 4.14.

The bright cloud in the lower half of Fig. 4.13 is a spectacular sight during the cold clear nights of the Australian winter. The first time I saw it from the darkness of Siding Spring mountain I thought it was a wisp of cloud caught in moonlight. But there was no Moon that night and what appears to be nebulosity is composed of vast multitudes of relatively faint stars that congregate around the nucleus of our Galaxy. They are the same kind of stars that we saw as the yellowish haze around the nucleus of M83 in Fig. 4.10. Though we see them clearly enough in external galaxies such as M83, in our Galaxy their appearance is still strongly affected by the veils and sheets and clouds of dust that gather between the stars.

The fact of this obscuration was well known to Walter Baade, who studied the

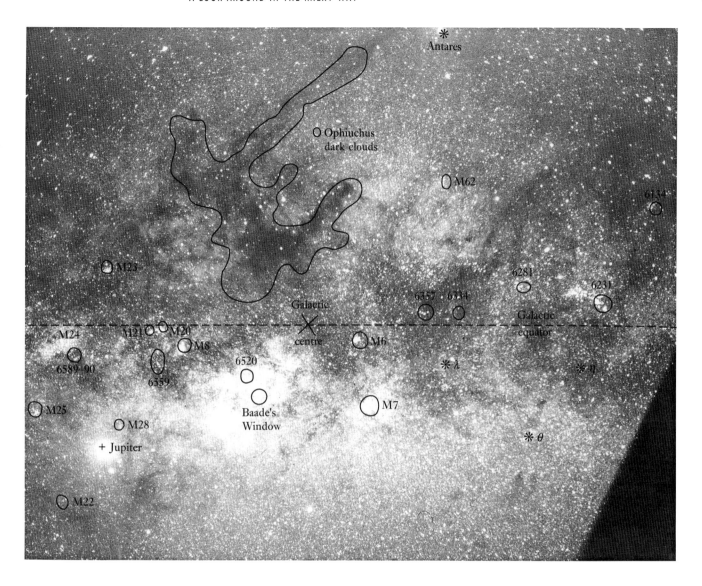

Antares

Ophiuchus
dark clouds

M62

6134

M23

6281

6231

6357  6334

Galactic
centre

Galactic
equator

M24    M21  M20

M8

M6

6589–90

6559

6520

✳ λ

✳ η

M25

Baade's
Window

M7

M28

✳ θ

+ Jupiter

M22

*Fig. 4.14.* Location map for the major objects in Fig. 4.12, the wide-angle picture of the Milky Way around the Galactic centre. NGC objects are identified by four digits.

*Fig. 4.13.* These two spectacular nebulae are almost on the Galactic equator. NGC 6334 and 6357 are partly hidden in the dust that diminishes the number of stars seen in this direction and changes the colour of the nebulae.

Milky Way extensively from the Mt Wilson Observatory in California during and after the Second World War. He was also the first to recognise clearly that there are two basic types of star in most spiral galaxies, including our own. There are those intrinsically faint, relatively cool stars that are found in huge numbers in the Galactic bulge, which he defined as Population II, and the much rarer, highly luminous stars of Population I that are found in the spiral arms. It was therefore Baade who recognised the different populations of stars in Gould's Belt and the Galactic bulge that we described earlier. Though this division into just two categories is now seen as an oversimplification, it is still valid and perfectly explains the range of subtle hues we see in colour pictures of spiral galaxies.

Among the Population II stars are those that vary in their light output, pulsating with a regular beat. Remarkably, the frequency of the pulsation is related to the true luminosity of the star. Generally, the more luminous (and thus more massive) the star, the slower is its regular swing between maximum and minimum light. Several kinds of

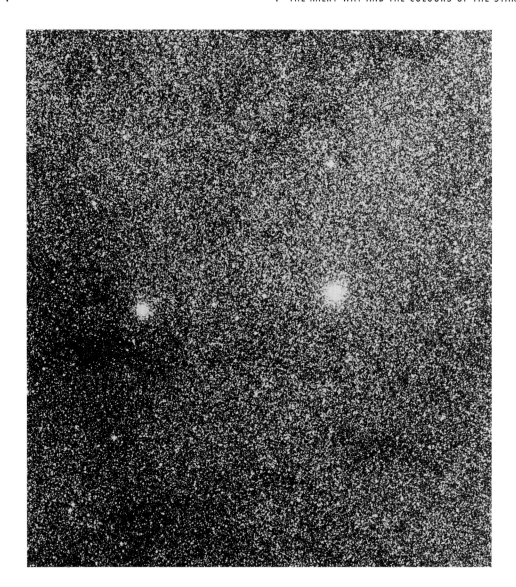

*Fig. 4.15.* Baade's Window is centred on the globular cluster NGC 6522 and offers a glimpse into the densely packed stars in the one of the brightest parts of the Milky Way. This region was studied by Walter Baade in the 1940s in his effort to measure the distance of the centre of the Galaxy.

such stars are known and they are of interest because they can be used to establish astronomy's elusive third dimension, distance. By comparing several photographs taken over a long period, Baade found a sprinkling of such variable stars in the hosts of old stars around the Galactic nucleus.

He made many exposures of a region in Sagittarius where he believed he could see deep into the star clouds of the Galactic bulge. By careful examination of wide-angle photographs, he found a small patch of sky where the obscuration seemed to be at a minimum, and was able to show that in this direction the number of variable stars first rose with distance and then fell. He was using the variables both as a tracer of the old stars swarming around the Galactic nucleus and as a distance indicator. The result showed that Baade had seen through the crowded bulge to the less crowded region on other side, and he was able to demonstrate that the distance to nucleus of the Galaxy was about 30 000 light years, which corresponded to the distance of the maximum number of vari-

*Fig. 4.16.* Irregular dust clouds hide the starry background surrounding Baade's Window (centre) in Sagittarius. The tiny dark cloud Barnard 86 is near the top of the picture, which covers almost the full 6.5 × 6.5 degree field of a UK Schmidt plate. The Galactic centre is in the direction of the heavily obscured region in the upper right of the picture.

*Fig. 4.17.* The star clouds of Sagittarius photographed in blue light by pioneer astronomical photographer Edward Emerson Barnard as part of a project to catalogue the dark markings in the Milky Way.

ables. More recent estimates put it at 27 000 light years; Baade had made a major contribution to understanding the size of the Galaxy.

The region that Baade selected for special study is centred on the globular cluster NGC 6522, which is itself at the centre of Fig. 4.15. This patch of sky is now known as Baade's Window. There are two globular clusters in the photograph; the second, NGC 6528, is to the left and its colour shows it to be partly obscured by dust. It is no accident that we see globular clusters here. These ancient, crowded cities of stars form another population of objects in orbit around the Galactic nucleus with paths that carry them far beyond the stars of the bulge and the components of the flat disk. The globular clusters and the stars of Baade's Window are thousands of light years away. In the near foreground of Fig. 4.15, about 120 light years distant is the bright orange, 'red' giant star Gamma Sagittarii.

A wider view towards the Galactic centre is seen in Fig. 4.16, which shows the location of Baade's Window in the Sagittarius star clouds. The complex nature of the

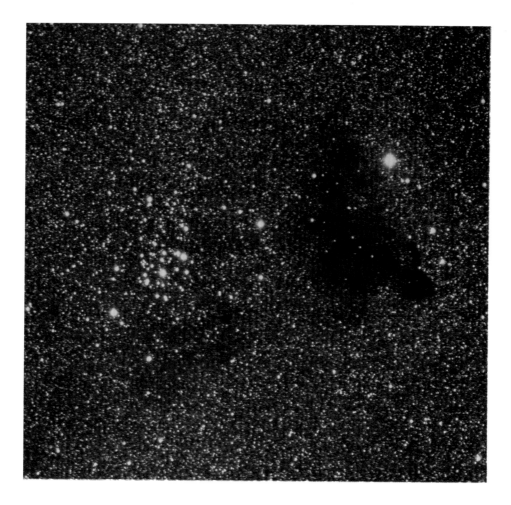

*Fig. 4.18.* A more detailed look at Barnard 86 and the nearby open cluster NGC 6520 is obtained with the superb optical quality and 1° field of the AAT. This picture shows both the colour of the stars and the fading and reddening of the background stars around the edges of the dark cloud of dust.

dust clouds is evident, as is the colour effect of increasing dust thickness. Where the stars appear fewest and faintest, on the right of the picture, they also seem yellowest. This, and the previous photograph, Fig. 4.15, were made from the same set of plates taken with the UK Schmidt Telescope. This telescope has a focal length of about 3 metres (120 inches) and provides a wide field of view, about 6° from top to bottom, and almost all of that angular coverage is seen in Fig. 4.16.

Towards the top of the picture is a tiny dark cloud, listed as number 86 in the catalogue of American astronomer Edward Emerson Barnard. Barnard was a self-taught astronomer-photographer who was fascinated by the dark spaces in the Milky Way, and whose photographic work did so much to establish their nature. He, too, used a wide-angle camera in his pioneering studies of the Milky Way. Initially, it was fitted with a 250 millimetre (10 inch) portrait lens, with even wider coverage than the UK Schmidt Telescope, and Barnard's photograph of this part of the sky is seen in Fig. 4.17. At the other end of the scale of focal length is a modern reflector such as the AAT, where angle of view is traded for much better resolution, quantity for quality. Fig. 4.18 shows Barnard 86 as seen in colour with the AAT, which has a focal length of 12.7 metres (500 inches). It was while taking the plates for this picture that I obtained the wide-angle view of the Milky Way that was used in Fig. 4.12.

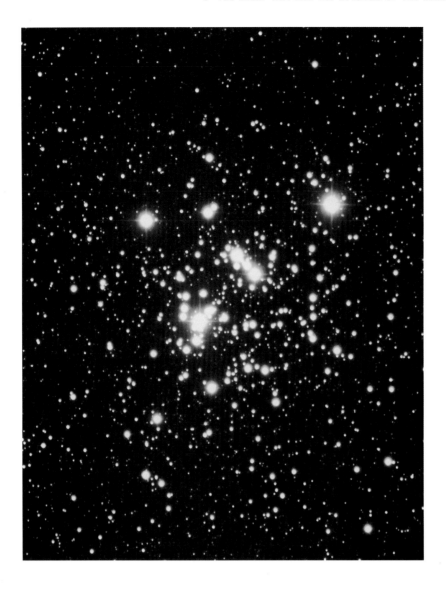

*Fig. 4.19.* The beautiful open cluster of stars NGC 4755, Sir John Herschel's famous Jewel Box in Crux.

Initially, Barnard, and the many astronomers who went before him, followed the ideas of William Herschel, a German musician turned astronomer who lived in England. At about the time of the American Declaration of Independence in 1776, Herschel was embarking on a career that would lead him to devote his life to unravelling 'the construction of the Heavens'. He was among the first to use a telescope to study the Milky Way in detail, and became the most influential astronomer of his day. Herschel believed that the dark holes he saw in the Milky Way were voids torn in its ancient starry fabric by the ravages of time, through which one could peer to the Universe beyond. Eventually, Herschel was shown to be wrong, but it was not until about 1930 that the final proof came.

In the colour photograph (Fig. 4.18) it can be seen that the dark space of Barnard 86 has an ill-defined edge where the stars seem to fade away. These stars are seen to be yellower as well as fainter around the periphery of the dark cloud, exactly what might be expected where light passes through the less dense extremities of a semi-opaque cloud of small particles. Unfortunately, colour pictures were not available to Barnard, nor super-

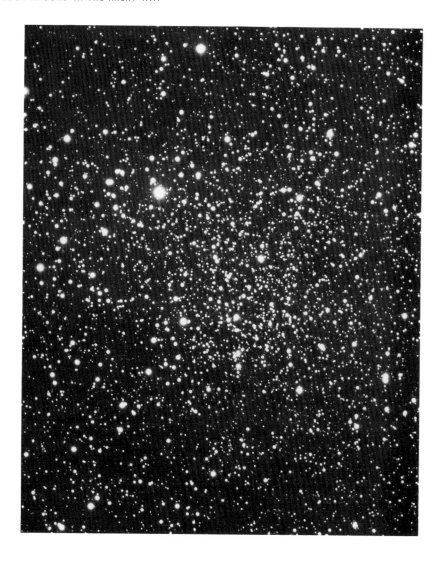

*Fig. 4.20.* Trumpler 5 is an old cluster of stars, almost indistinguishable from its starry surroundings. It is devoid of the bright stars that distinguish young star clusters. (See Fig. 6.19).

human night-time colour vision to Herschel, which is why the explanation we accept today took so long to be acknowledged.

In the same field of view is an open cluster of stars, NGC 6520, the brighter members of the group revealing in their markedly different colours the presence of young, massive stars at various stages of their evolution. Most are blue, with a few orange-yellow ones scattered among them. Many such open clusters of widely differing ages can be found sprinkled throughout the Milky Way. It is from the study of the colours of the stars in clusters of this kind, where the members of the group are all at the same distance and have formed at about the same time and from the same materials, that astronomers learn about stellar evolution. Fortunately, not all are seen against the crowded Sagittarius star-fields. Like Barnard 86, NGC 6520 is in the foreground, and the dust cloud may well be the dark remnants of the molecular cloud that gave birth to the cluster of stars several million years ago.

An open cluster that shows the diversity of stars that it contains is better seen against a less crowded background in Fig. 4.19. This is the Jewel Box, so named by

William Herschel's son John, also a famous astronomer. He noted it during his pioneering survey of the southern sky, which he made from South Africa in the 1830s, and described the cluster in his usual fulsome way. 'The stars which compose it, seen in a telescope of diameter large enough to enable the colours to be distinguished, have the effect of a casket of variously coloured precious stones.'

Wherever large terrestrial 'precious stones' are found in their natural environment they are usually heavily outnumbered by smaller, less eye-catching examples. So it is with stars. While there is a mere handful of the brightest stars, the fainter ones are much more common, and around them, perhaps more numerous still, there may be lesser bodies, planets. But all the stars in the group will have formed at about the same time, and by measuring the colours of a large number of them the age of the cluster can be determined. In the case of the Jewel Box, this turns out to be about 15 million years.

In a few tens of millions of years, the short-lived bright stars of the Jewel Box will run their course, turning from blue to yellow-orange and eventually exploding as supernovae. During their 'blue' period these energetic, massive stars eject a substantial fraction of their substance from their searingly hot surfaces. During their much cooler red supergiant phase they produce prodigious amounts of tiny particles which drift away from the stars like smoke. The influence of this dust from the stars has had an effect on our view of the Universe out of all proportion to its mass, which is a minute fraction of that of the Galaxy.

The intense radiation and vigorous outpourings of these short-lived stars is not conducive to the evolution of life on any planets they may have, and the final supernova may in any case destroy them. If the cluster contains only a few hundred stars, the less explosive, lightweight stars that remain after 100 million years may drift away, and, as the group evaporates, the cluster will lose its identity. In a more massive cluster, such as the ancient group Trumpler 5 (Fig. 4.20), the stars may remain together for much longer. Either way, the fainter members, perhaps among them some like the Sun, will live on for billions of years, perhaps long enough for their planets to develop life intelligent enough to look into the sky and wonder at what it is that they see.

# 5 ASTRONOMICAL REFLECTIONS

## INTRODUCTION

'Astronomical reflections' might be used as the title for a biographical piece by some fading astronomical luminary, an amiable ramble down memory lane. Instead, in this chapter we will look at an aspect of optical astronomy that has not been much explored in a way that is easily accessible. The reflections I refer to are reflection nebulae, a fascinating and interestingly assorted class of objects that tell of the presence of dark matter in space of which we might otherwise be unaware.

The first searchers of the skies who were equipped with telescopes saw extended, hazy-looking objects between the stars, which they identified as 'nebulae' (the Latin word for mists or clouds). Even before the telescope was invented, the Andromeda 'nebula' and the Magellanic Clouds had been seen with the unaided eye as faint, permanent and mysterious patches of light. By far the biggest and best-known example of an extended or nebulous night-time object was the Milky Way. That too was a long-standing mystery.

But eventually, these fuzzy features of the night sky were sifted into two distinct categories. There are those objects that emit light because they are very hot, and glow by virtue of the temperature of the atoms of which they are composed. The best celestial examples of this kind of thermal radiation are of course the Sun and the stars, and the things that are largely made of stars such as the Milky Way, other galaxies and globular clusters.

In a telescope all but the nearest galaxies and many globular clusters look like largely formless, feebly glowing patches of light. When their light is split into its component colours by a prism or a diffraction grating it is spread into an almost unbroken range of rainbow colours and is described as a continuum spectrum. This light has the same character as sunlight and the best Earth-bound example of this kind of light source is a tungsten filament lamp.

Unfortunately, there are also celestial objects that appear to be extended in the telescope, like globular clusters and galaxies, but that are themselves not especially hot. They glow because they have been excited into emitting light by absorbing energetic but invisible radiation from something nearby that is very hot. This radiation is then re-emitted in a form that we can see, rather like the light from a domestic fluorescent tube. These objects are emission nebulae, and although they might look like galaxies and distant clusters of stars in a telescope, they have spectra that are quite distinctive, consisting of a series of very narrow lines of pure colours.

Both emission nebulae and galaxies glow in the dark; they can be seen and photographed. But it was the invention in the 1850s of the spectroscope, a device for analysing the spectral composition of light, that showed them to be fundamentally different. The spectroscope and its modern equivalent, the spectrograph, provided the key to understanding the composition of the distant Universe, and their early application to astronomy led to the important distinction between galaxies of stars and luminous clouds of gas.

A third class of extended object that glows in the dark, but that is *not* intrinsically luminous, are the reflection nebulae, and since they can have the continuous spectra of stars but are often associated with emission nebulae it is easy to understand why they were confused for so long. The definition of reflection nebulae can also include objects that predominantly scatter light rather than simply reflect it, and, surprisingly, this can include things that one might normally think of as perfectly transparent, such as gases.

The best example of a reflection nebula lit by starlight is the daytime blue of Earth's atmosphere. We do not normally think of it as such because it is so familiar, but that is merely because most of us do not think of such everyday sights in terms of the world of physics and astronomy. The atoms and molecules of the air are not especially efficient light scatterers, but there are a lot of them. However, as pointed out in Chapter 2, they are much more efficient at scattering the shorter wavelengths of sunlight, which is why the sky is blue.

Air molecules are small, a few ten-thousandths of a micrometre (micron) across. A micrometre is itself pretty small, one thousandth of a millimetre; there are about 25 millimetres to the inch. To give some sense of scale to this tiny world, the middle range of visible light has a wavelength of about half a micrometre. The solid particles that scatter blue light efficiently are significantly smaller than that, between 0.01 and 0.2 micrometres in diameter, much bigger than air molecules and typical of the size of particles in smoke. We see smoke rising in air as distinctly blue, though there are many fewer smoke particles than air molecules along our line of sight. It is thus smoke-like particles that are the main scatterers of light in interstellar space.

'Smoke' between the stars hides most of the Milky Way from view, and often changes the apparent colour of those parts that are not hidden. This same finely divided material, illuminated by the light of stars, produces some astonishingly beautiful scenes, especially when associated with other kinds of nebulae, and since reflection nebulae are usually found mixed in with other kinds, the pictures in this chapter cannot be exclusively devoted to objects that reflect light. In many parts of the Milky Way all these effects can be seen in the same photograph.

## DUST, GAS AND REFLECTION NEBULAE

In the previous chapter we saw the star clouds of Sagittarius, criss-crossed by lanes, clumps and veils of dust. In Fig. 5.1, I present another wide-angle view of the Milky Way. It reveals subtle changes in the colour of the background stars caused by

Fig. 5.1. The bright Omega nebula, M17 is in the upper part of this wide-angle UK Schmidt view of the Milky Way at the northern boundary of Sagittarius. Much fainter are two small, blue patches of reflection nebulosity, NGC 6589–90 near the lower edge of the photograph. They are seen in more detail in Fig. 5.2. Across the rich field of stars thin veils of dust affect the colours of those in the background and in places hide them completely.

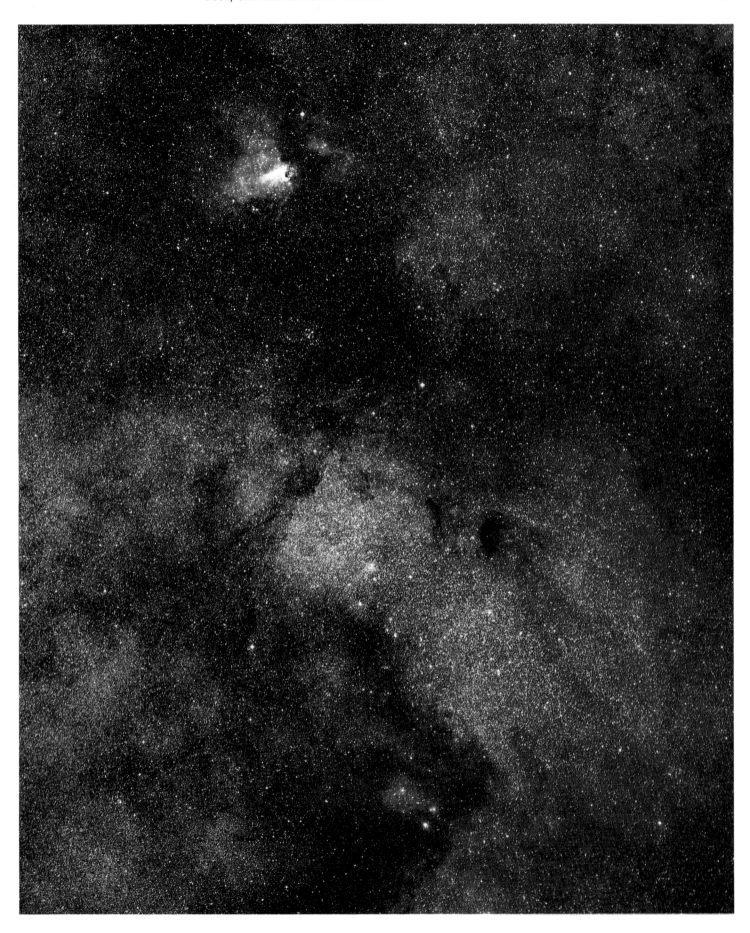

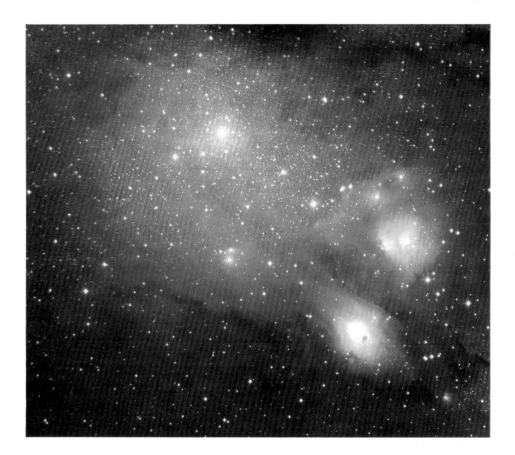

*Fig. 5.2.* With the AAT, the blue blobs of NGC 6589–90 can be seen as the brightest parts of a large, faint reflection nebula that merges with the emission nebula IC 1283–84, diluting its fierce red to a softer magenta hue.

irregular interstellar absorption across almost 6° of sky, a very wide-angle view by astronomical standards but typical of images from the UK Schmidt Telescope. At the top of the picture is a bright red star-forming region M17, itself intimately associated with dust.

Towards the bottom of the illustration, in a region of fewer stars, is a less spectacular emission nebula that Messier probably could not see and certainly did not catalogue. This is the complex seen in more detail in a picture made from AAT plates, Fig. 5.2. It contains the two blue reflection nebulae that appear in a catalogue much later than that of Messier, the *New General Catalogue* of 1888, the source of all the NGC numbers scattered throughout this book.

These two nebulae, listed in the NGC as 6589 and 6590, surround stars that have probably drifted into the dust that spreads throughout the region. The stars themselves are hot, much hotter than the Sun. Although they are not hot enough to generate enough ultraviolet light to make a large emission nebula, they are sufficiently luminous for their light to be scattered over an extensive region by the tiny grains of dust. Throughout the photograph the 'smoke' is nowhere completely dark; it is feebly reflecting light from other nearby stars, and from the Milky Way itself.

Towards the centre of the picture, the dust thins so that we are able to see more of the background stars that Fig. 5.1 suggests might be there, and the traces of hydrogen that remain are illuminated by the brightest star in the picture. This star is hot enough to

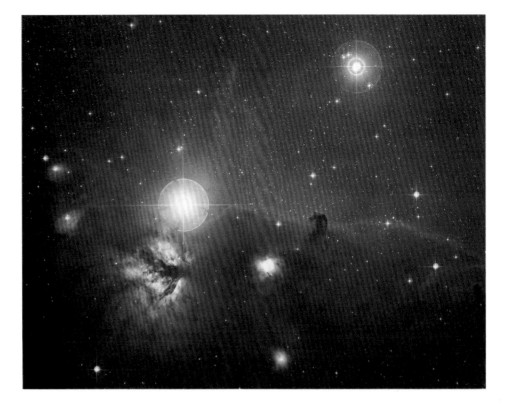

*Fig. 5.3.* At least five small reflection nebulae are scattered around the Horsehead nebula in Orion. The largest of these is NGC 2023, at the base of the horse-head shape. Their blue colour is light scattered from stars caught up in the outskirts of dusty clouds, seen elsewhere as dark nebulae, especially as the famous Horsehead nebula itself, silhouetted against a nebulous background. The large, yellow nebula, partly hidden in the glare of Zeta Orionis, is NGC 2024.

excite an emission nebula, and the scattered blue light and red fluorescence blend to a soft shade of magenta. In nebulae such as this we see evidence of some of the many ways that dust can scatter, absorb and reflect the light from stars, and how it can combine with emission nebulae to produce a wide range of colours.

Perhaps the most memorable image created by the interplay of starlight, gas and dust is that of the Horsehead nebula in Orion. A wide-angle view of this is shown in Fig. 5.3. The Horsehead itself is a partially opaque projection from a much larger dark cloud that is illuminated by Sigma Orionis, the bright star at the top right of Fig. 5.3. The dark cloud is dense, and because starlight penetrates it hardly at all, very cold within. Deep inside, protected from destructive photons of ultraviolet light from the hot stars, the chilled dust particles become coated with traces of water ice, solid carbon dioxide and other simple molecules that are gases or liquids at room temperature. These compounds are made from the lost outer atmospheres of massive, cool stars, long vanished.

Over countless centuries, the mixture is occasionally stirred by a penetrating cosmic ray. The chemically active materials on the dust particles gradually become more complex, forming sizeable organic molecules. Elsewhere in the dark cloud, lone hydrogen atoms use the solid surfaces offered by the dust grains to fuse into pairs, creating molecules of hydrogen gas. This is a molecular cloud. The molecules emit weakly at radio wavelengths, each species at a characteristic frequency, announcing their chilled existence and something of their nature to radio astronomers.

If there were no nearby stars, this molecular cloud would be just another entry on the radio astronomer's large inventory of interesting sources scattered darkly throughout

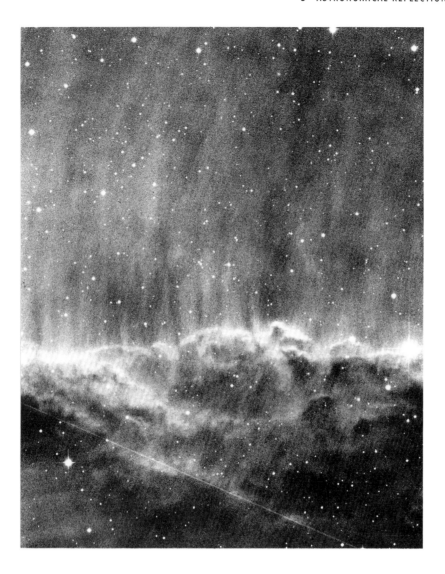

*Fig. 5.4.* The irregular edge of the molecular cloud from which the Horsehead nebula emerges is IC 434. It is being destroyed by intense radiation from a nearby star, releasing curtains of glowing gas. The straight line across the picture is the trail from an artificial satellite.

the Milky Way. However, in Orion there is little chance of invisibility. This constellation is both nearby and filled with brilliant stars, many of which are at about the same distance as the dark clouds, about 1500 light years away.

One of the hottest of the stars in Orion is Sigma Orionis. Sigma's ultraviolet radiation is eroding the molecular cloud, destroying delicate chemical bonds and releasing hydrogen atoms. The surface which is being eroded appears as a narrow strip of vivid red against which the horse-head shape is seen. This is catalogued as IC 434. In a region to the right (south) of the Horsehead, a specially-enhanced black and white picture (Fig. 5.4) reveals that the irregular, heated surface of the molecular cloud is releasing gas from the dust in long streamers.

To the left of the Horsehead in Fig. 5.3 is the grossly over-exposed image of Zeta Orionis, a star bright enough to have a proper name, Alnitak. Though the three-dimensional distribution of the stars in this direction is not known with any accuracy, it seems that Zeta is in the foreground and plays only a small part in illuminating the nebula, which is probably several hundred light years beyond. However, its swollen image and cross-like diffraction spikes, which are telescope artefacts, partly hide the curious yellowish nebula NGC 2024, about which surprisingly little is known. What is known has been learned only in the last few years and with the aid of telescopes and detectors capable of working in the infrared and millimetre-wave part of the spectrum, detecting radiation that penetrates dust much better than the longest wavelengths we can see. These techniques show the cold gas to be clumpy, containing warmer condensations which may be the first stages of star formation.

This well-known region also contains a number of blue reflection nebulae, the most conspicuous being NGC 2023, near the base of the Horsehead. The size of the image of Zeta has been reduced on this picture, and the images 'sharpened', by a photographic process called 'unsharp masking'. This image was the first UK Schmidt picture that I made in colour and is among the best known. Its success encouraged me to make the many others that appear in these pages.

## INTERSTELLAR DE-BLUEING

The visible nebulosity that characterizes hydrogen emission seen in the photographs of the Horsehead nebula is not a pure red, though red is the dominant colour. As well as the red 'H alpha' line of excited hydrogen, weaker colours are present, most prominently the 'H beta' line of hydrogen, which appears in the blue part of the spectrum. Thus the colour of this emission nebula is a mixture, mainly of red, but with a blue component, i.e. a shade of magenta. As we have seen in Chapter 4, the tiny, smoke-like particles that comprise interstellar dust filter out blue light, so where there is emission nebulosity seen through dust its colour is changed from magenta to a fainter but purer red by selective absorption. This effect can be seen in Fig. 5.5 in the lower jaw and mane of the horse-head shape, where the dust does not completely hide the emission nebula beyond.

Two other manifestations of dust are seen in this picture. The blue reflection nebula at the bottom of the frame is NGC 2023. It is of interest because it has been found to contain polycyclic aromatic hydrocarbons – organic molecules – which are not normally identifiable in visible light. Such delicate molecules are found only near stars that are too cool to destroy them, but hot enough to expose them. Also here is a group of what seem to be tiny red worms burrowing their way through the dark cloud. These are Herbig–Haro objects, often the first visible signs of activity from newly-formed stars still hidden in dust. They are often found in star-forming regions such as this because young stars are sometimes unstable as they settle down to a steady energy production and occasionally eject rapidly moving plumes of the material that falls onto them from the surrounding cloud. These jets travel through the dust at very high velocity and their interaction with their surroundings produces small, quite distinctive, wispy nebulae. Both the filaments of molecular material and the wispy Herbig–Haro jets are seen in the negative print in Fig. 5.6. Both were discovered on the AAT plates that I took to make the colour picture in Fig. 5.5.

In some parts of the Galaxy the dark clouds of dust are hardly illuminated at all by nearby stars. Such clouds appear in silhouette against the starry background, like those in

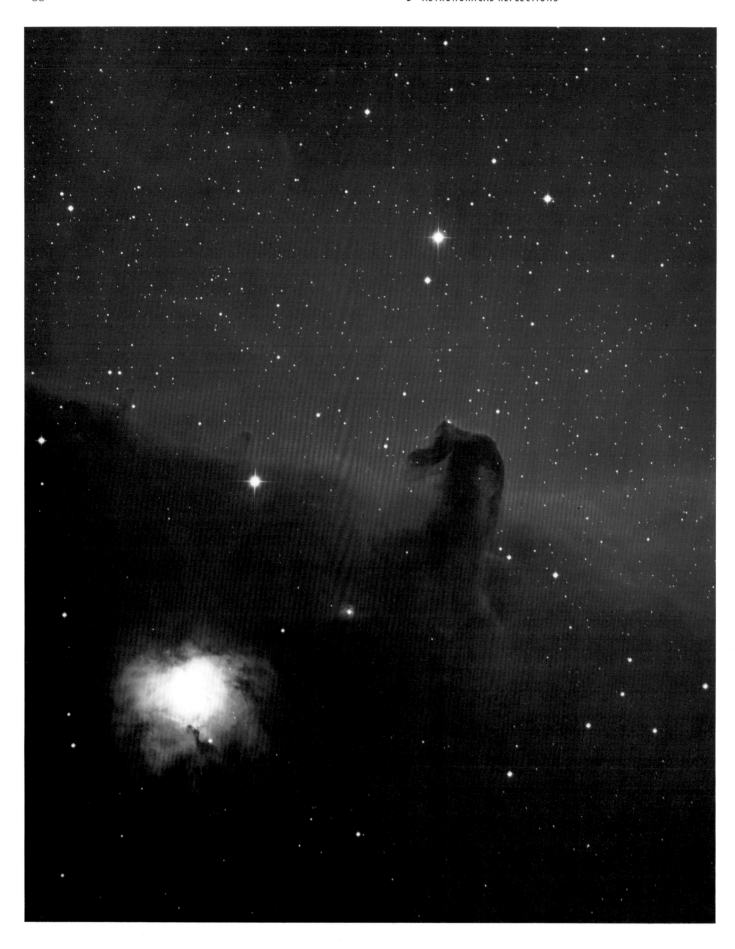

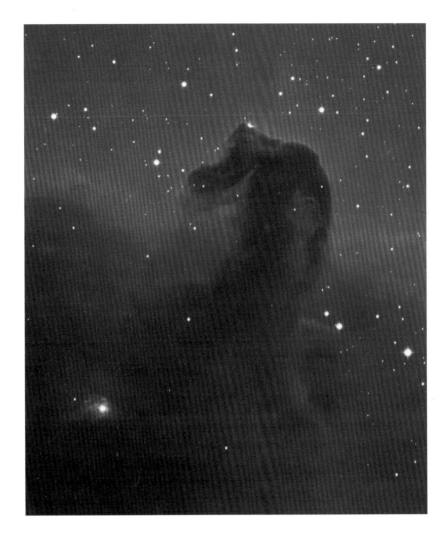

*Fig.* 5.6. A closer look at the Horsehead nebula itself shows the colour effects produced by absorption of light by dust. In the 'lower jaw' and 'mane' of the horse-head shape the magenta nebulosity is filtered to a duskier red by the thin dust clouds. This is interstellar reddening of an already red nebula.

*Fig.* 5.5. The 'neck' of the Horsehead nebula is a dusty protrusion from a large molecular cloud, parts of which are lit up by embedded bright stars producing blue reflection nebulae. The red nebula itself shades gradually to magenta near the edge of the molecular cloud where more of the light from Sigma Orionis (not shown) is scattered by dust particles released as it is destroyed.

the complex emission region near NGC 6188 (Fig. 5.7). As near the Horsehead, some patches of dust are illuminated by starlight so as to produce distinctive blue reflection nebulae.

In some places in the Galaxy whole nebulae are seen through dust. The two faint objects that were used to define the Galactic plane in Fig. 4.13 are pictured in more detail in Figs. 5.8 (a) and (b). Unlike the vivid magenta shades and subtle textures in the Horsehead nebula and the soft blue shades in NGC 6188, these two star-forming regions seem curiously flat. They are deep in the dust of the Milky Way and are about 5500 light years away.

Most photographs of emission nebulae show that the red nebulosity surrounds one or more very hot, bright blueish stars; these same stars produce the ultraviolet radiation that splits the hydrogen atoms apart. In NGC 6334 and 6357 there is no evidence of a blue component in their colour – indeed, the blue-light plate used in this three-colour picture had no nebulous image at all – nor is there any sign of the bright blue stars normally found in these objects. The line of blue stars scattered across the face of NGC 6357

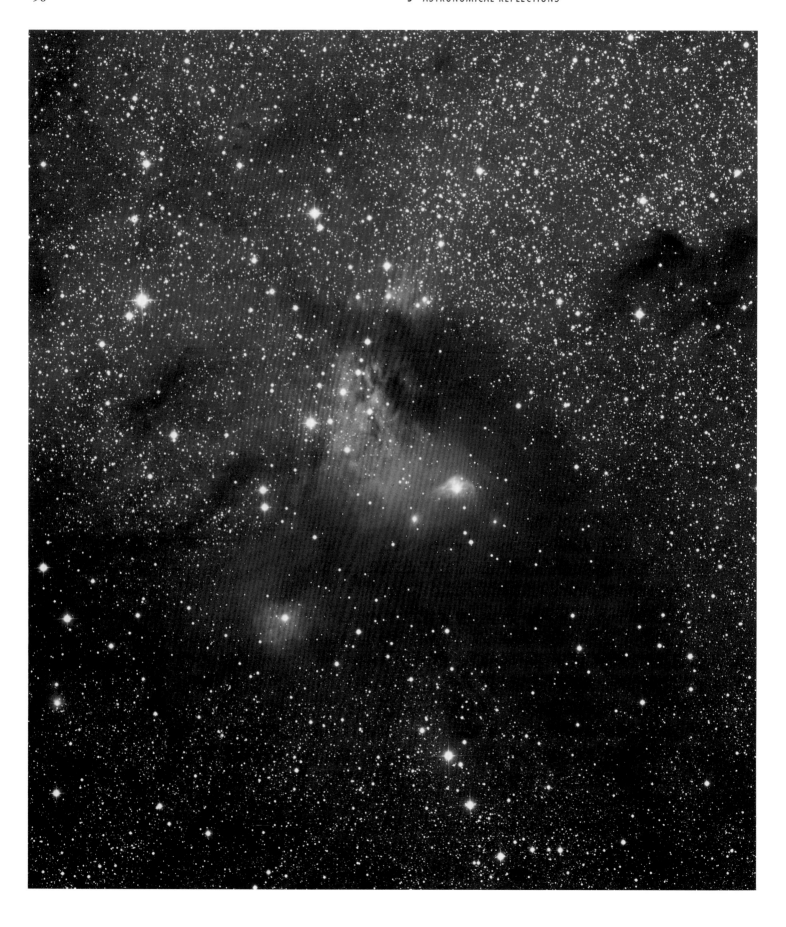

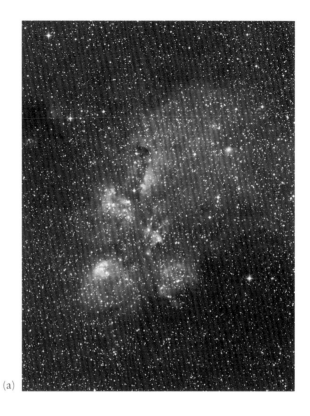

(a)

*Fig.* 5.8. (a) and (b)   NGC 6334 and NGC 6357 are star-forming regions in Scorpius. They are buried in the dust that gathers around the plane of the Galactic equator and most of their blue and green light and that of the stars that excite them is obscured.

*Fig.* 5.7.   Soft blue reflection nebulae are seen when a bright star illuminates part of a dust cloud in the emission nebula NGC 6188. Dust that is not lit by starlight is seen as dark against the rich starry background.

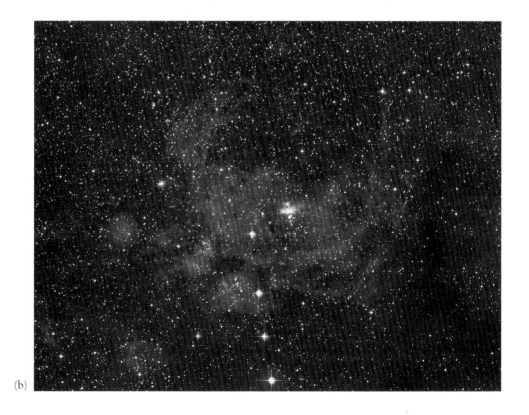

(b)

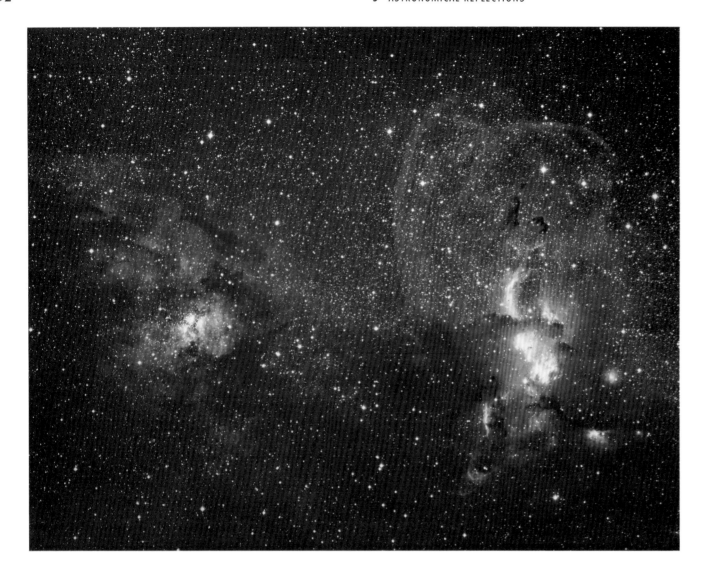

(Fig. 5.8(b)) are in the foreground and play no part in creating the nebula. These nebulae are excellent examples of de-blueing (interstellar reddening), the selective removal of blue light by particles of dust in the line of sight that so strongly affected our view of the Milky Way in the last chapter. This wavelength-dependent absorption accounts for both the ruddy hue and apparent absence of blue stars. The stars are present, but only part of their light penetrates the dust and so they are not conspicuous. Careful measurement of their colour shows that the stars and their associated nebulae are dimmed by a factor of about 10 in the green part of the spectrum, even more in the blue region, but relatively little at red wavelengths.

Interstellar reddening in a red nebula is demonstrated even more dramatically in Fig. 5.9. Though these two objects appear side by side on the sky, the nebulosity associated with the remarkable cluster of stars NGC 3603 (left) is on the far side of the Carina spiral arm of the Galaxy at a distance of about 20 000 light years, and is seen through considerable amounts of dust. Its apparent neighbour, the curious looped nebula NGC

*Fig. 5.9.* The colour difference between these two nebulae is mainly due to absorption of blue light, and it indicates that the star cluster NGC 3063 (left) is more distant than the looped nebula NGC 3576 that seems to be joined to it.

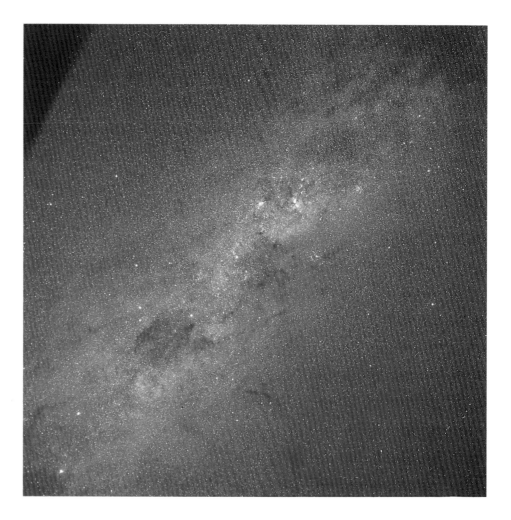

*Fig. 5.10.* A wide-angle view of the constellations of Crux and Carina includes the Southern Cross and the Coalsack and takes in the most spectacular stretch of the southern sky.

3576, is less than half as far away and is on 'our' side of the Carina arm, which is the westward extension of the dust clouds in Sagittarius that blocked our view in Fig. 4.12. NGC 3576, on the other hand, is hardly obscured at all and the strong effect of interstellar absorption on the appearance of these objects is obvious from the colour pictures, though undetectable in black and white.

While the plates for Fig. 5.9 were being taken, I had attached a Hasselblad camera loaded with colour film to the prime focus cage of the AAT, peering out through the dome in the same direction. The shutter was open during the 90 minutes or so it took to make the three colour exposures, and the prime focus cage was not moved. The wide-angle picture that it took covers over 30° of the sky and it is shown as Fig. 5.10. In the centre, almost invisible, are the two nebulae that are so striking in Fig. 5.9, which covers a little less than 1° of the field. That they are quite insignificant is hardly surprising. The AAT mirror gathers over 20 000 times as much light as the lens of the Hasselblad, and there is a similar factor in their cost.

Fig. 5.11 identifies the objects in this fascinating part of the southern sky, which includes parts of Centaurus as well as Carina and Crux, around the Southern Cross. Alongside the Cross is one of the nearest of the dark clouds that hide so much of the

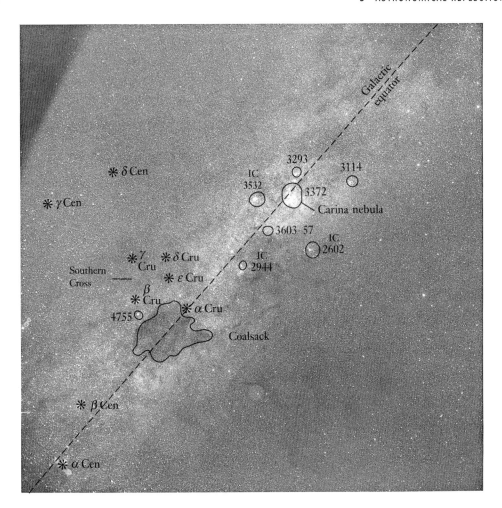

*Fig. 5.11.* The principal stars and nebulae in Fig. 5.10 are identified. The four-digit numbers identify NGC objects. Despite the richness of this part of the sky, there are no Messier numbers here because it was too far south to be seen from Paris.

Galaxy from our gaze. It is a prominent naked-eye object to southern observers, and is nowadays known as the Coalsack. According to Bart Bok, among much else the author of a famous book on the Milky Way, some Australian aborigines saw the Coalsack as the head of an emu whose body and neck stretched the whole length of the southern galaxy, expanding into wings and feathers in Sagittarius and with spindly legs and splayed feet reaching into Scutum.

The dark hole of the Coalsack is all the more prominent because it lies alongside the brilliant asterism of the Southern Cross and is a dark intrusion into the generally bright background of Gould's Belt. The Coalsack cannot be seen from the city, nor is it visible in moonlight. On the best nights in the darkest places it appears as an inky blackness containing a single sixth magnitude star on the threshold of visibility. This star is a good test even for an experienced and dark-adapted eye. The Coalsack is also used by southern star-gazers, both amateur and professional, as an informal measure of the darkness and transparency of the night sky. It is not too far from the south celestial pole and is thus almost always visible from the walk-way of the AAT dome for a quick assessment of observing conditions during the night. Though the long-exposure photograph shows

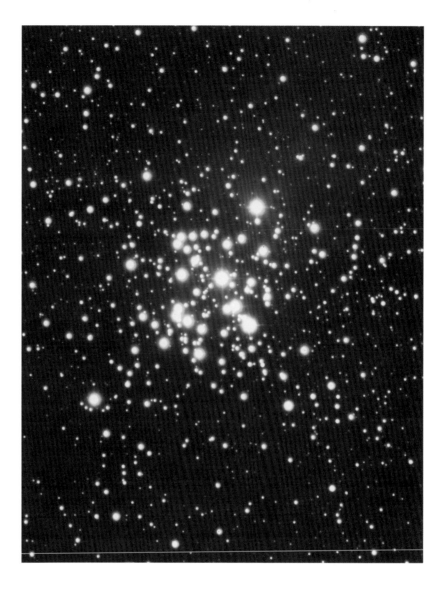

*Fig. 5.12.* Faint wisps of nebulosity still linger around the young open cluster NGC 3923. One of the young blue stars has turned into a red giant. Despite the name, red giants appear orange-yellow on colour photographs, and to the eye.

many stars within its blackness that are much fainter than can be seen with the unaided eye, the Coalsack is only obvious because of the many more stars that it hides completely.

Fig. 5.10 also reveals numerous open clusters of stars scattered across the field. Some are visible to the unaided eye, like Kappa Crucis alongside the Southern Cross. Others come to life in binoculars or a modest telescope, but in a large telescope huge numbers of them appear. Just beyond the striking Carina nebula, the biggest red blob in Fig. 5.10, the inner spiral arm that we first encountered almost face-on in Sagittarius curves away from us out of sight. In the direction of Carina we are therefore looking at the tangential point in the spiral arm and into a great concentration of those young objects that define the outer edges of spiral arms, as seen in the spiral galaxy M83.

Some of the clusters of stars are so young that they are still associated with the gas and dust that produced them. Traces of this can still be seen as the faint red emission and

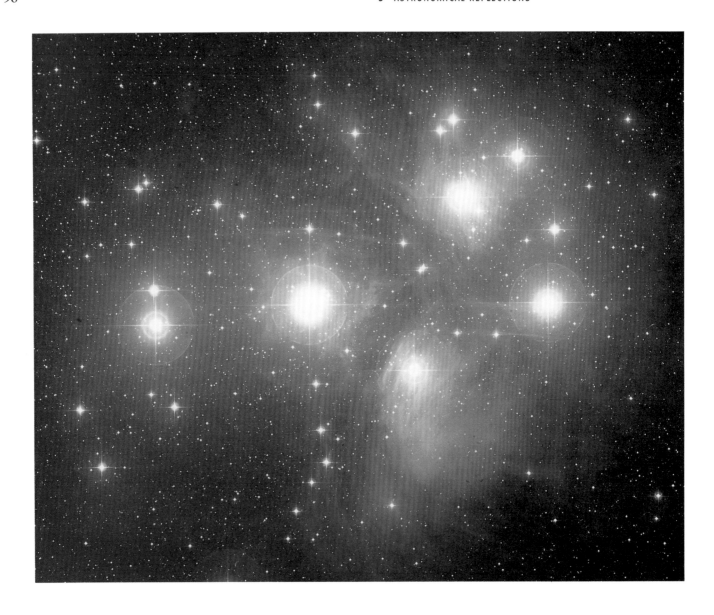

blue reflections in Fig. 5.12 near the open cluster NGC 3923, which is about 10 million years old. This group of young stars is also in the rich constellation of Carina, and is at a similar stage in its development to the Jewel Box, illustrated in Fig. 4.19.

Not all reflection nebulosity associated with open clusters is the remains of the material that gave rise to the stars. In the case of the the Seven Sisters, perhaps better known as the Pleiades (Fig. 5.13), this sparkling group of naked-eye stars seems to have drifted into the tenuous dusty outskirts of a molecular cloud. The wispy 'combed hair' structures are not associated with the stars themselves, but are the result of the alignment in the interstellar magnetic field of the tiny particles responsible for scattering the light. This effect was probably present long before the stars arrived to reveal it.

The blue colour of the reflection nebula is Rayleigh scattering of starlight by small particles. One of the bright Pleiades, Merope (lower left), is the closest to an invisible molecular cloud, whose presence has been detected by radio observations. This star seems to

*Fig. 5.13.* The best-known reflection nebula is associated with an equally well known group of stars, the Pleiades, or Seven Sisters. The streaky appearance is typical of dusty interstellar material that is largely unaffected by its illuminating stars.

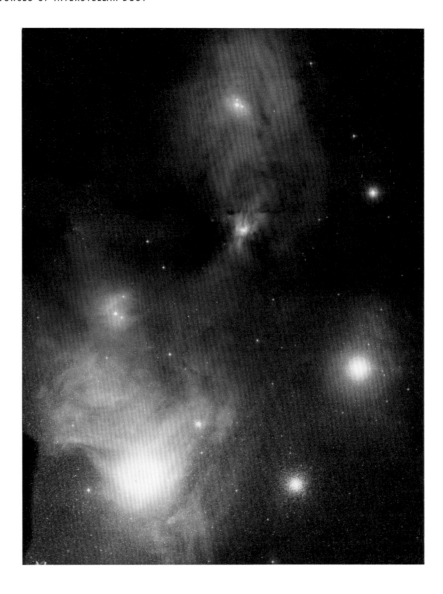

*Fig. 5.14.* The nebulae associated with the hot star Rho Ophiuchi (upper centre) is blue, while the much cooler Antares (lower left) is immersed in a yellow cloud that it has created itself. Almost all of the astonishing range of colours seen here is from starlight reflected by dust particles.

be illuminating particles more numerous and perhaps larger than those elsewhere. The reflected light is therefore more like that of the star from which it came and much less blue than the rest of the nebula. Once again, the colour image reveals a subtle effect that is not visible in the black and white images of this part of the sky.

## SOURCES OF INTERSTELLAR DUST

There are many sources of the smoke that deflects light travelling between the stars of the Milky Way. Probably the most prolific producers of small particles are the massive, cool, red supergiant stars such as Antares, seen at lower left in Fig. 5.14. In this picture, the star image seems washed out, hidden in a reflection nebula of its own making. The star itself is better seen in Fig. 1.16(a), an exposure made on colour film. Its outer envelope is so cool, a relatively modest 3000 K, that refractory materials such as silicates

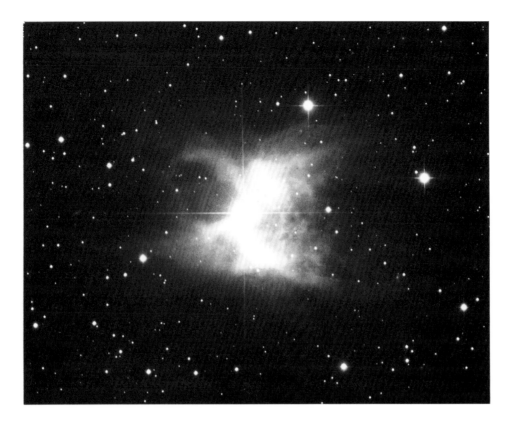

*Fig. 5.15.* The Toby Jug nebula, IC 2220, surrounds a bright but cool star and is the result of light reflected from particles that the star itself has ejected.

and metal oxides are able to condense out of the elements created by nuclear fusion inside the star. Other stars produce carbon (soot) and carbides and nitrides, but examples of cool stars with visible nebulae are quite rare. Fig. 5.15 shows another, the IC 2220 nebula, popularly known as the Toby Jug nebula from its likeness to an English drinking vessel.

The smoky atmosphere of the star further reddens the light seeping through from within. Eventually, as they drift away from the star, the small dust grains will be pushed by stellar winds or pulled by gravity to gather into the cold, dark clouds like those seen elsewhere in Fig. 5.14. Where the dust is thick enough to mask its interior from energetic photons, the heat is radiated away from the gathering clouds and they cool, so that volatile materials – water, carbon dioxide, ammonia – condense on the surfaces of the solid particles. It is in such places that slow, low-temperature chemical reactions create surprisingly complex organic molecules.

As in the Pleiades, occasionally stars drift by to illuminate the outskirts of the molecular clouds, producing the distinctive 'shot silk' appearance seen in reflection nebulae that have not been disturbed by the illuminating stars. This effect is often best seen in a black and white image such as Fig. 5.16, which was specially printed with an unsharp mask to bring out the subtle structures around the Rho Ophiuchi group of stars. These stars are seen to be surrounded by a blue nebula in the upper part of Fig. 5.14, the Rho Ophiuchi reflection nebula.

Here and there this light is mixed with emission nebulosity or attenuated by yet

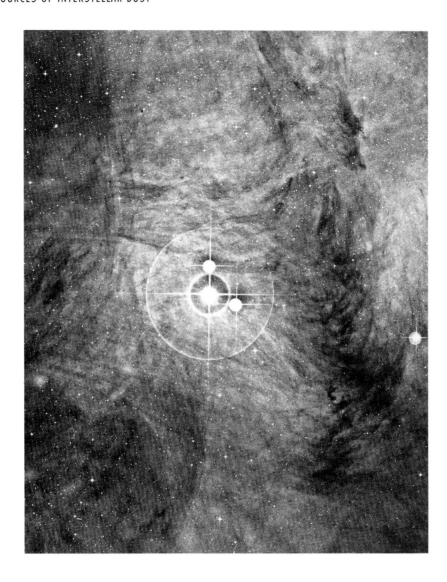

*Fig. 5.16.* The 'combed hair' streak-iness of the reflection nebula around Rho Ophiuchi is reminiscent of that around the Pleiades (Fig. 5.13) and is revealed by unsharp masking a deep, blue-light plate from the UK Schmidt Telescope.

more dust, producing a surprising range of colours. But behind the colourful veils can be seen dark clouds, pregnant with young stars not yet visible. It was this part of the sky that William Herschel described as a hole in the heavens, a void through which he believed he could see the Universe beyond. His eyes were not sensitive enough to respond to the nebulosity that we now know fills the field. What he saw was not an empty, distant Universe, but a nearby stellar nursery, for the moment merely reflecting starlight, but soon to produce its own. The nebulosity here is truly faint; its apparent intensity has been increased by another process, photographic amplification, which I applied to the original UK Schmidt plates from which Fig. 5.14 was made.

Reflection nebulae are often faint because the particles that are involved are few and far between and are, in any case, inefficient reflectors. Usually they are seen to reflect only starlight, but recently one example has been found which seems to be a reflection nebula reflecting the light from another nebula. In Fig. 5.17 we see the yellowish glow of the Orion nebula reflected in a nearby dusty cloud (NGC 1973–75–77) associated with an

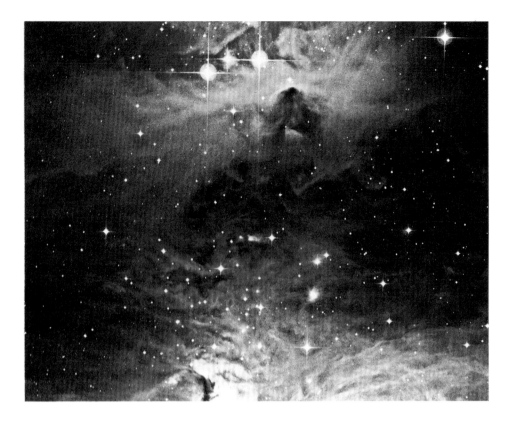

*Fig. 5.17.* An extreme example of photographic manipulation reveals the yellowish glow from the Orion nebula (off the bottom of the picture) reflected in the dusty underside of the nebulosity associated with the NGC 1973–75–77 cluster of stars.

open cluster of stars. So why is this reflected light not blue like other reflection nebulae? There are two possible answers, and both are concerned with the source of the light that has been reflected.

The Orion nebula, discussed more fully in Chapter 6, is itself a dusty nebula and light from the stars within it, as seen from the NGC 1977 region, is probably stripped of its blue component by dust. It could be this filtered starlight that we see reflected in NGC 1977. More probably, it is the combined red (hydrogen) and green (oxygen) emission from the brilliant core of the Orion nebula itself. Green and red light mix to produce yellow, but only spectroscopic examination of the faint yellow reflection will decide its origin. While black and white pictures of this scene reveal its complexity, they give no hint that the illuminating source of the 'underside' of the NGC 1977 nebula is another nebula. It should be mentioned here that I used unsharp masking in combination with photographic amplification to produce Fig. 5.17, exaggerating the colour contrast and emphasising the fine detail, so the colour balance is likely to be distorted.

A more 'normal' view of NGC 1973–75–77 is seen in Fig. 5.18. If it were suspended anywhere else in the sky, this nebula would be regarded as a magnificent spectacle. However, this group of stars and its nebulosity is just half a degree – the apparent size of the full Moon – north of the much brighter Orion nebula and has largely been ignored because of it. This group of hot stars looks to the unaided eye like a single object, the northernmost 'star' of the sword handle of Orion, the group of three

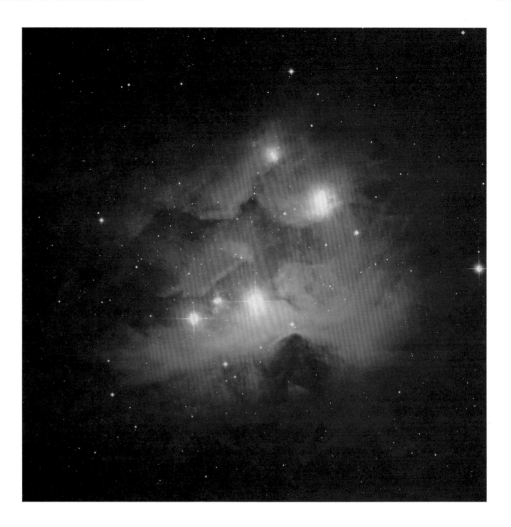

*Fig. 5.18.* The nebulosity around the NGC 1973–75–77 cluster in Orion.

labelled '16' in Fig. 4.5. Most of the nebulosity is reflected light from this cluster of stars. However, they seem to be in a cavity, perhaps created by outflows from the stars themselves. Within it, wisps of hydrogen fluoresce red, lending something akin to a warm inner glow to a cool reflection nebula. Together, the stars and nebula create an impression of quiet beauty quite unlike the maelstrom of the adjacent Orion nebula.

There is one important reflection nebula whose interest lies in the fact that it is no longer there. It has never been photographed because it disappeared in the 1830s, at about the time photography was invented and long before the new-fangled technique was sufficiently sensitive to record nebulae. However, though long suspected by visual observers, the disappearance of the nebula was finally established by photography in the 1890s. The photographs were taken from Sydney Observatory, by the Government Astronomer of the day, Henry C. Russell. The object in question was a part of that spectacular jewel of the southern Milky Way, the nebula in Carina.

The region was sketched in considerable detail by Sir John Herschel from his temporary observatory in Cape Town, South Africa. He was there from 1834 until 1838, sur-

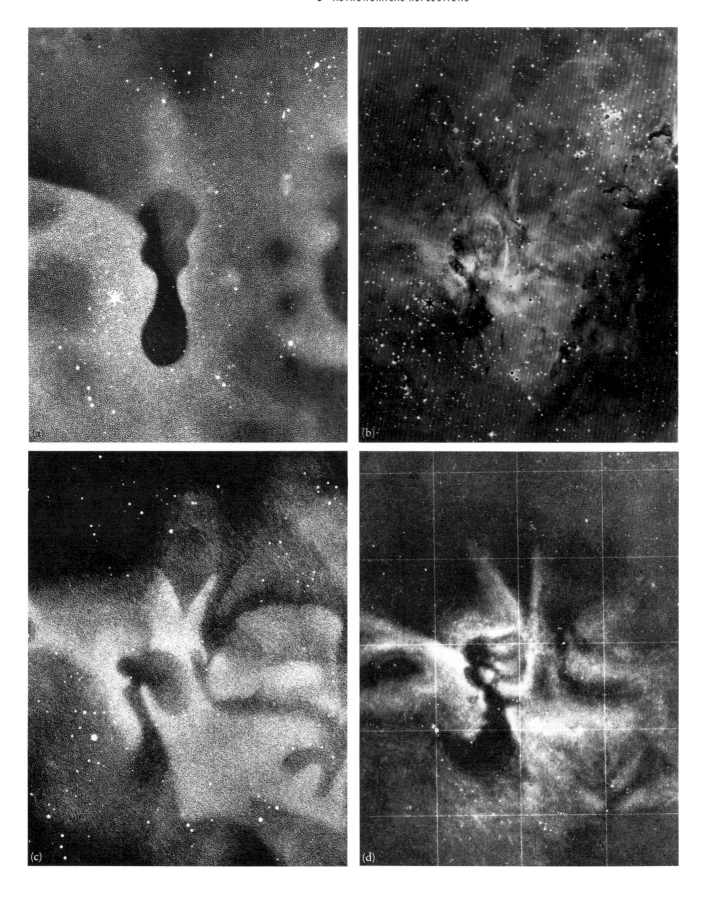

veying the southern skies with as much diligence as his father, the famous William, had surveyed the north. While he was at work in Cape Town, the star Eta Carinae, which had a long history of variability, brightened from insignificance until it was among the brightest in the sky. Herschel's drawing is reproduced as Fig. 5.19(a): it shows Eta as the most prominent star.

Herschel particularly noted the bright rim to the west (right) of this upstart star, and, for obvious reasons, called the whole dark region the Keyhole nebula. He was convinced that what he saw was an ephemeral phenomenon and took a good deal of care with his drawings to ensure that he included sufficient stars (which could be located exactly on the sky) to define more precisely the position of the nebula (which could not). In the years after Herschel left the Cape, Eta once again faded into obscurity, but interest in it has remained very much alive.

One of the reasons that was given for installing the 48 inch Great Melbourne Telescope in that southern Australian city in 1867 was to compare what Herschel had seen in the 1830s with later drawings made with a bigger instrument. Some of the best Great Melbourne Telescope sketches were made by Fairie MacGeorge in 1870. His rendition of the Eta Carinae region (Fig. 5.19(c)) shows that the eastern side of the Keyhole, between Eta and the dark space, had vanished. Drawings made by another of the telescope's observers at about the same time (Fig. 5.19(d)) confirmed MacGeorge's impression. But a close look at the drawings reveals that some of the features of the nebula that we now know have not changed significantly are rendered differently. Such are the difficulties in comparing drawings that the reality of MacGeorge's discovery remained in doubt.

The fate of the southern rim of the Keyhole nebula was not confirmed until it failed to appear on Russell's 1890 photograph, in the same way that it fails to appear on modern photographs. In Fig. 5.19(b) I have superimposed Herschel's drawing (as a negative, with black stars) on a recent AAT colour picture. It was a curious feeling combining my recent photograph with Herschel's sketch made over many nights at the eyepiece 160 years ago. One wonders what this great astronomer would have made of this once-familiar ground if he could see it now in colour. Most of the bright stars in the two images match as well as can be expected, given that some of the them will have moved in the years that separate the two, but there is little trace of the southern (lower) part of the Keyhole that was so prominently bright to Herschel.

In all probability the brightening star illuminated the rim of a dust cloud alongside, and what Herschel saw was a reflection of Eta's light, which suggests that the star and nebula are closely associated in space. A faint trace of the reflection nebula can still be seen in the subtle colour differences near the star's elongated image in Fig. 5.20. The reasons for Eta's unpredictable behaviour and its other visible manifestations are discussed in Chapter 7.

Solid particles between the stars are commonplace, but, as we have seen, are very difficult to detect at visible wavelengths unless there is some special geometry or unusual lighting effect. The recent supernova in the Large Magellanic Cloud (LMC), provided both. The appearance of a supernova in the Milky Way's nearest

*Fig. 5.19.* (a) Sir John Herschel's drawing of the Keyhole nebula and Eta Carinae, published in 1847. In (b) Sir John Herschel's drawing of the nebula is overlaid on a modern colour photograph. Many of the stars that do not match exactly may have a 'proper motion', a real shift in the position of stars that is easily seen in the 150 years that has elapsed between the pictures. In (c) is reproduced a sketch of the Eta Carinae region made by Fairie MacGeorge with the Great Melbourne Telescope in December 1870. (d) Le Sueur's drawing of Eta Carinae made with the telescope in January 1870. Both show that the southern part of the 'keyhole' had vanished since Herschel's time.

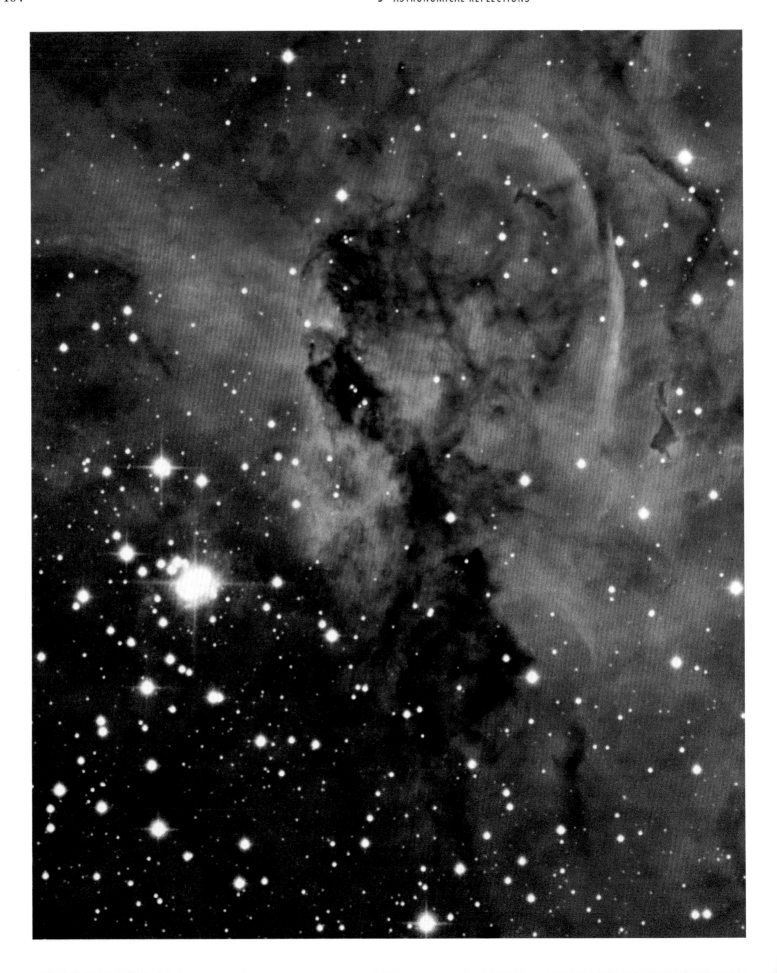

companion galaxy was completely unexpected, though not unanticipated. Astronomers have speculated for years about what they would do if a nearby supernova were to appear. After all, none had been visible to the unaided eye for almost 400 years.

Supernovae are important to astronomers for many reasons, especially to those who speculate about what happens inside stars. The only possibility of checking on their speculations, upon which so much of our knowledge of the workings of stars is based, is to study those that reveal their inner secrets when they explode. Unfortunately, much of the elaborate theory of stellar structure and evolution had been tested only against distant – and thus faint – examples. The arrival of light from the supernova in the nearby LMC sent a wave of excitement through the astronomical community that is still detectable.

Apart from its importance in testing theories of stellar evolution, the LMC supernova provided a splendid, probably once-in-a-lifetime chance to detect solid particles between the stars and to map their three-dimensional distribution in space. The brilliant flash of the supernova was detected in February 1987. The supernova reached maximum brightness in May of that year but has now faded so that it is very much fainter than the star that exploded. This brief, intense flash of light travelled for over 170 000 years before being detected on Earth. However, some of the light has been deflected from its original direction by tiny dust particles near, but not on our line of sight to the supernova. This light has travelled slightly further to reach us and its arrival has thus been delayed: the more distant from the line of sight, the longer the delay.

The dusty regions encountered are in the form of two distinct concentrations of particles 400 and 1000 or so light years in front of the supernova. The dust, itself the detritus from massive stars, has probably been swept up by earlier generations of stellar outbursts. But the line of sight of the supernova is also filled with nebulosity and the problem of detecting new faint structures against this complicated background was considerable. Fortunately, I had a three-colour set of plates of this part of the sky that were taken before the supernova appeared, and by a simple (in principle!) photographic subtraction process I discovered that the feeble light echo image could be extracted from them, recombined and revealed – in colour.

Reflections from the two dusty veils are seen as two concentric circular light echoes in Fig. 5.21, which was made from plates taken in February 1989, subtracted from otherwise identical, pre-supernova plates of the same region taken in 1984. The colour accurately (and literally) reflects the yellowish hue of the supernova at its brightest, in May 1987, and is thus a true echo of its light. The same colour is seen in Fig. 7.22, a direct photograph of the supernova.

In between the main rings of the light echo, smaller clouds of scattering particles are seen. New blobs of dust appear as the light echo expands through the interstellar medium of the LMC and a time series of photographs of the kind shown in Fig. 5.22 can be used to map the distribution of dust increasingly distant from the line of sight. The expansion of the rings between July 1988 and January 1992 is apparent. I take new plates of this faint echo at every opportunity, though now I must add three like images together

*Fig. 5.20.* The peculiar star Eta Carinae is itself surrounded by a small, red nebula and has faded since Herschel's time, but it still illuminates a faint blueish reflection nebula near what was once the prominent southern extension of the Keyhole nebula.

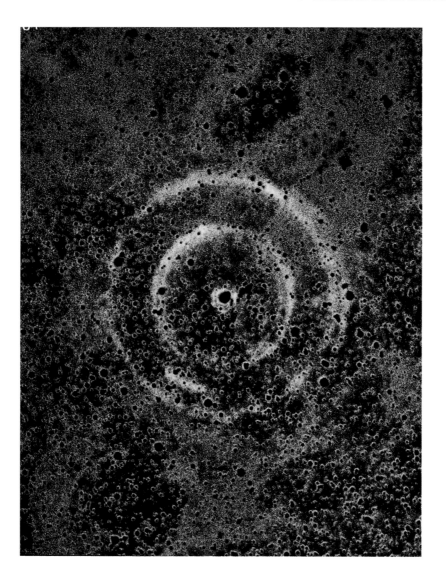

*Fig. 5.21.* As the flash of light from supernova 1987A moves outwards from the exploding star it is scattered by two sheet-like regions of otherwise invisible dust, arriving long after the supernova (centre) has faded. The faint, scattered light is seen later than the supernova that produced it and appears as a light echo. The yellow colour in the rings is that of the supernova in May 1987 when it was at its brightest. The picture was made by a subtraction technique that produces black images of the brighter stars and greenish hues in some of the nebulosity where image cancellation is incomplete.

to strengthen the image as it fades. Eventually, we should have enough separate pictures to make a short time-lapse movie that shows light from the supernova spreading out between the stars.

The geometry of the situation is surprisingly complicated. As seen from Earth, the rings are illuminated sections of a very long but narrow ellipsoid, with the supernova at one of its foci and the observer at the other. A sketch of the relative positions of the various components is reproduced as Fig. 5.23.

The last reflection nebulae we will look at will provide a link with the next chapter, on the visible manifestations of the birth of stars. The Trifid nebula, M20, has long been the most familiar example of a dusty blue nebula associated with a beautiful red emission region. Photographs of it appear on the next few pages. Here I display a new reflection feature, discovered by colour photography using deep plates from the telescopes of the Anglo-Australian Observatory.

*Fig. 5.22.* The expansion of the light echo is revealed by subtracting a pre-supernova image from a series of more recent exposures taken at irregular intervals over the last four years. The first picture was taken in 1988, the last in 1992.

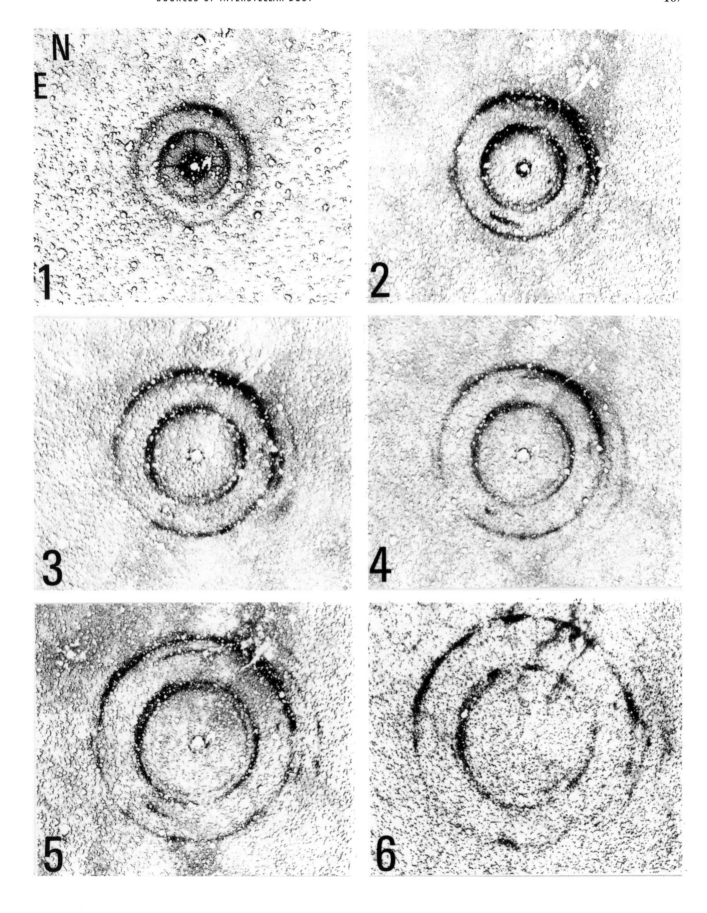

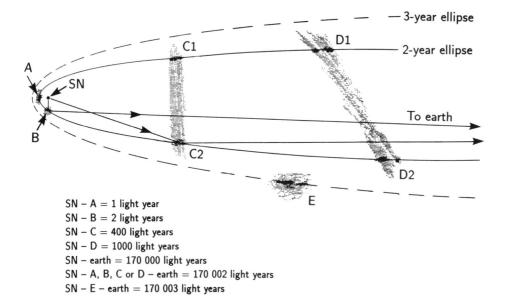

SN – A = 1 light year
SN – B = 2 light years
SN – C = 400 light years
SN – D = 1000 light years
SN – earth = 170 000 light years
SN – A, B, C or D – earth = 170 002 light years
SN – E – earth = 170 003 light years

*Fig. 5.23.* The geometry of the light echo of supernova 1987A. The echoes are reflections from dust caught in the sphere of light emitted by the supernova when it was very bright. As seen from Earth, this sphere traces a very elongated but expanding ellipse with the Earth and the supernova at its two foci. The two- and three-year ellipses and are sketched (not to scale) and the location of the sheets of dust that produce the rings of the light echo is shown diagrammatically.

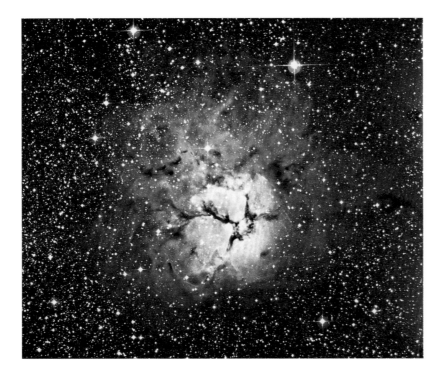

*Fig. 5.24.* M20, the Trifid nebula, is surrounded by a faint reflection nebula that was discovered on this picture, which was made by combining unsharp masking and photographic amplification techniques on plates taken by the UK Schmidt Telescope.

In Fig. 5.24, we can see not only the well-known northern reflection nebula around a bright star, but another, fainter example completely surrounding the emission region, which is roughly spherical. Short-wavelength, invisible ultraviolet light from the stars within is completely absorbed by the large amounts of hydrogen in this part of the Galaxy, producing the red glow. But the edges of the red nebula do not mark the limit of the hydrogen; away from its source, the ultraviolet component of the light from the central stars is gradually depleted by absorption by the gas and the absorption of

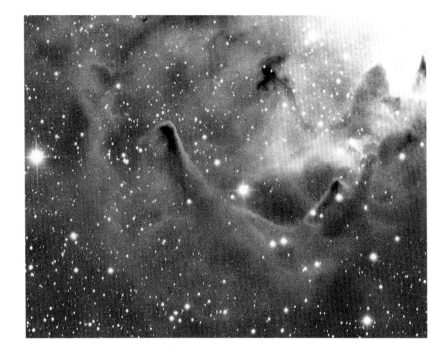

*Fig. 5.25.* With the AAT, we can see in more detail a part of the boundary where the red emission region of the Trifid nebula changes into blue reflective cloud, marking the edge of the sphere of ultraviolet light from the central stars shown in Fig. 6.3.

ultraviolet light thus ultimately defines the extent of the nebula. However, beyond the red emission region sufficient blue light remains to be scattered by the surrounding dust grains. A closer look at the region where the emission nebula fades and the reflection nebula begins (Fig. 5.25) reveals the complexity of this interface.

Another intriguing aspect of reflection nebulae has also been revealed for the first time by three-colour photography of the Trifid nebula. This is the subtle greenish hue seen when a blue reflection nebula is partly obscured behind an absorbing cloud. As we have seen, the dust strongly absorbs blue light, absorbs green a little and red hardly at all. The light of the reflection nebula, on the other hand, is mostly blue, with some green and hardly any red, thus the colour of a reflection nebula most likely to penetrate the dust is green, and that is what we see in Fig. 5.26, a detail of the western part of the Trifid nebula.

As this chapter was being finalised, I took some new plates of two very faint reflection nebulae that the UK Schmidt deep sky survey had suggested might be interesting if I could extract a colour image. The objects were indeed faint, in one case so much so that the resulting colour picture was 'noisy', but the colours revealed are astonishing and quite unobtainable by any other process.

One of these nebulae is associated with a small dust cloud with a wispy tail, an object known as a cometary globule. This one is simply listed in the sparse literature of these objects as CG 4 and appears in Fig. 5.27. It is an excellent example of an isolated dust cloud being destroyed by radiation from nearby, energetic stars. We see the trail of dust swept away from the dark cloud by unseen stars off the eastern (left) side of the photograph. The tail is blue because it is a normal reflection nebula. Where light from the tail is seen through the dust, near the head of the globule, greenish hues are seen, shading to yellow-brown here and there, as in the Trifid.

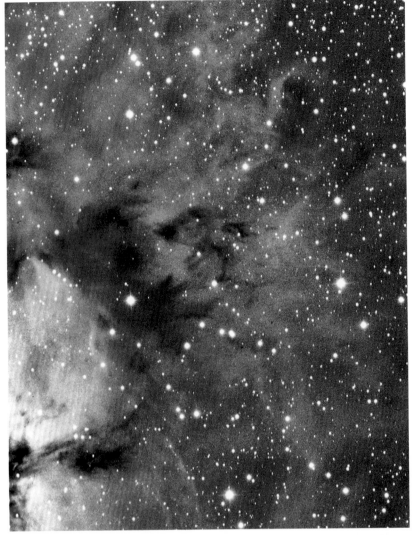

Fig. 5.26. Where the reflection nebulosity is seen through dust at the edge of the Trifid nebula most of its blue light is absorbed, leaving a fainter, greenish-yellow hue.

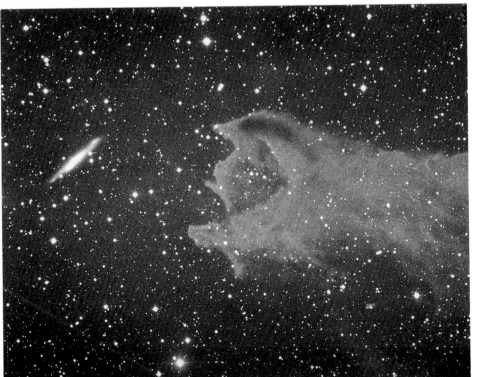

Fig. 5.27. The colourful head of the very faint cometary globule CG 4 is surrounded by a diffuse red glow of hydrogen liberated from it by the radiation of nearby stars. These same stars are probably responsible for creating the dusty tail. As in the Trifid nebula, a greenish colour is seen where background reflection nebulosity is partly absorbed by foreground dust.

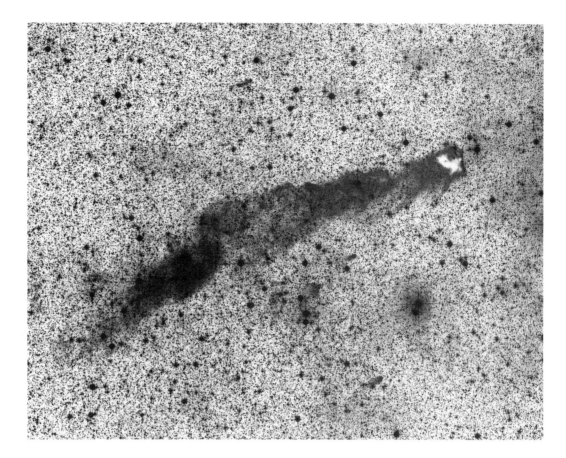

*Fig. 5.28.* Image enhancement techniques used with fine-grain, high-contrast sky survey plates turn up large numbers of faint cometary globules and isolated reflection nebulae. This example is CG 22, among the most spectacular.

As surprising as the green coloration is the faint red haze. This is almost certainly hydrogen, released from the dust cloud by bombardment with light from a surrounding star or stars, though why the hydrogen is not swept along with the dust tail is still a mystery. The most likely source of the energy that is eating away at CG 4 is Zeta Puppis. This is not a small-scale event. CG 4 is several light years across and contains enough material to make 10 or so stars with the mass of the Sun.

Many of these faint cometary globules are found on the new, deep photographic sky surveys now being undertaken. One of the longest globules is CG 22, shown in Fig. 5.28, and it is obvious from this picture how such objects received their generic name. Unfortunately for astronomical taxonomers, not all isolated dark clouds with bright rims can be swept neatly into the cometary globule category.

In Fig. 5.29 I publish the first colour picture of another kind of cometary globule, bigger and much more complex than CG 4, but similarly affected by radiation from stars off the picture. It, too, has a faint halo of red light around it but also some embedded stars that produce a few reflection nebulae within it. I can find no mention of it in catalogues of either dark clouds or of cometary globules, but the energy that is destroying it comes from a cluster of stars that is catalogued as NGC 6321. It is at the heart of a brilliant cluster of very young stars in Scorpius, well known to astronomers as the Sco OB1 association.

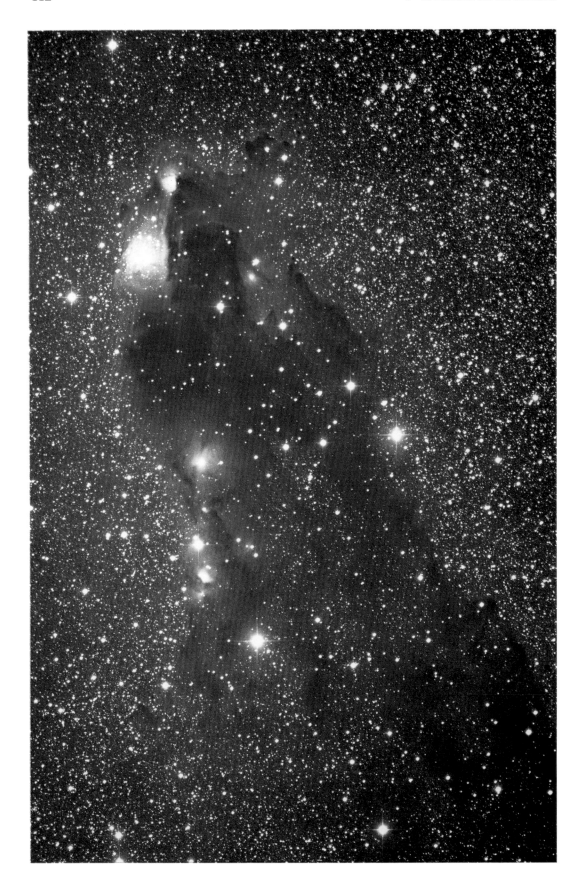

*Fig. 5.29.* A large cometary globule in Scorpius swept by stellar winds from a group of bright stars in the well-known Sco OB1 association.

*Fig. 5.30.* Bright stars embedded in dust associated with the Corona Australis nebula produce the two reflection nebulae NGC 6726–27 and are seen here for the first time in colour. Buried in the dust is a star formation that has given the dark cloud a faint yellow glow and produced a number of small, red nebulae.

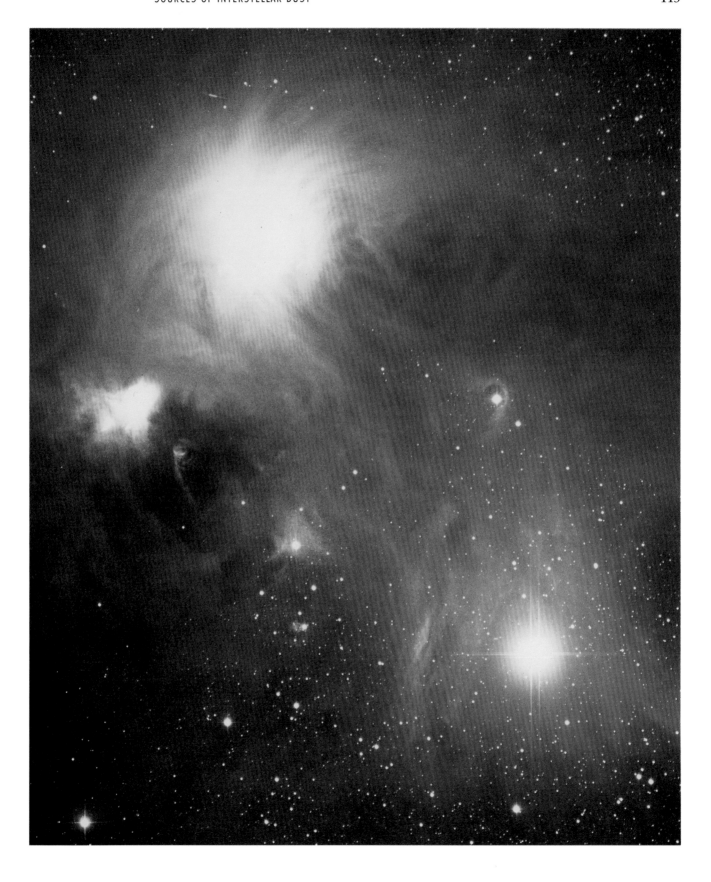

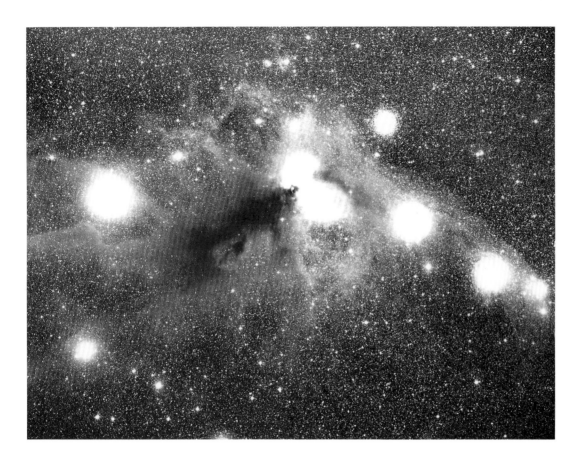

*Fig. 5.31.* The bright clouds seen in Fig. 5.30 are surrounded by a much larger faint reflection nebula revealed by combining images from several plates. The globular cluster is NGC 6723.

Another well-known group of stars lights up a small cloud of dust, 500 light years away in the Corona Australis region, producing the beautiful reflection nebulae NGC 6726–27 seen in Fig. 5.30. This is the brightest part of a very large but faint nebula that extends over many square degrees of sky (Fig. 5.31). But this large nebula, which contains enough material to make 130 Sun-like stars, is not a simple conjunction of bright stars and dust. In Fig. 5.30 at least two small, well-defined red nebulae can be seen, probably jets emitted from still-hidden stars in the dust cloud. These nebulae glow not by reflection of the light from the stars but by absorbing their energy and re-emitting it in characteristic colours. This mechanism is fluorescence, and in the next chapter it dominates where stars are formed, and although astronomical reflections are often present, they only serve to temper the brilliant hues of luminous gas.

# 6 WHERE STARS ARE BORN

## INTRODUCTION

As we have seen, the birth of stars occurs in cold clouds of gas, dust and molecules. We looked at some typical dark clouds illuminated by starlight in the last chapter. The gestation and early life of stars is played out within these dark places and is largely hidden from human eyes, even when aided by the biggest telescopes and the most sensitive detectors of light.

Fortunately, where there is action there is usually some heat, which leaves the dust as infrared radiation, which is why the dust stays cold. Nowadays, this radiation can be detected at infrared wavelengths. But the amount of heat seeping out from stellar nurseries is small, and until recently there was no useful imaging infrared detector for astronomy. This, and the fact that the telescope itself is an infrared radiator, makes infrared astronomy one of the most difficult and challenging branches of optical astronomy. So, despite major technological advances, as with photography 100 years ago, there are still many puzzles to solve about the places where stars are born.

One fundamental aspect of star formation that is still not well understood is the trigger; just what causes a molecular cloud to begin the collapse that leads to the creation of stars, and how does the collapse of a single cloud lead to the enormous range of stars, large and small, that are found in a young group? Astronomers pose such difficult questions because they are naturally curious people and enjoy a challenge. But this is not mere idle curiosity; the answers will tell not only of the evolution of distant groups of stars but also about the beginnings of the Sun and the planets around it, and, ultimately, about the conditions in the early solar system that led to the appearance of life.

Paradoxically, part of the answer to the riddle of star birth is likely to involve the vast amounts of energy released when stars die. Shock waves from supernovae may well provide the impetus for the compression and collapse of dusty clouds into new stars, especially since such explosions are most frequent in youthful, massive stars that are still associated with their birth material. During their short but vigorous pre-supernova lives, these same stars often expend much energy in generating powerful stellar winds that can also shock and compress any nearby gas and dust.

It seems that once started, usually by a few, massive stars, star formation is often self-propagating, and continues to spread through the gathered raw materials like a sporadic bush fire, until the gas and dust has been embodied in new stars or is dispersed by the vigour and violence of the star-forming activity itself.

The process of star formation may also be started by a gravitational interaction or collision between galaxies rich in gas. When galaxies encounter one another, which is not an unusual occurrence, the billions of stars that populate these stellar cities hardly ever meet face to face. In proportion to their size, the space between the stars is vast, whereas the galaxies themselves are both much bigger and relatively closer together. This is one reason why galactic meetings are common and stellar collisions are not.

But when gas clouds meet they tend to stick, and as they collide the injection of energy from their combined velocities can be sufficient to provoke vigorous star formation. Even without the impetus of a collision, gravitational interaction alone can stir quiescent molecular clouds into action. The best observable example of this is in the Magellanic Clouds, the nearest galaxies to our own. The Clouds are so close that they are interacting with the Milky Way and the dramatic effects of this inter-galactic encounter are discussed and illustrated in Chapter 8.

Even without outside influence, spiral galaxies such as the Milky Way are naturally efficient star-making factories because the necessary ingredients are squeezed by gravity into a narrow band in the plane of the spiral arms. Thus constrained, a relatively small additional disturbance, such as shock waves from supernovae, will promote collapse and the formation of stars.

As well as triggering star formation in such galaxies, supernovae may also destroy and disperse some molecular clouds, sometimes enough to halt star formation completely. The torn fragments of the molecular clouds can be blown high out of the plane of the galaxy, like vast bubbles bursting from molten lava. But the debris will soon begin to fall back again, attracted by the mass of the galactic plane, and it may return with a surprisingly high velocity. Such fast-moving clouds may themselves initiate star formation as they crash into the relatively stationary material trapped in the Galactic disk. This large-scale turbulence in the Galaxy is another direct consequence of the formation of massive stars, and it scatters their debris throughout the Milky Way. Traces of dust blasted from the disks of spiral galaxies can be seen in edge-on systems such as NGC 891 (Fig. 4.11) and NGC 3628 (Fig. 8.40). The heavy elements formed in supernovae are thus distributed as dust throughout a typical spiral galaxy, enriching the interstellar medium in elements other than hydrogen and helium.

Most young stars are found in clusters or groups which contain a few brilliant, massive luminaries and many lesser bodies. The ratio of lightweight to massive stars seems to vary according to the composition of the dark cloud, and seems to be sensitive to the concentration of heavy elements that it contains, amongst other possibilities. In the LMC, for example, where the abundance of elements heavier than hydrogen and helium is low, more massive stars seem to be produced. In the Milky Way, which has been greatly enriched in these 'metals' (oxygen, carbon, nitrogen etc. are 'metals' to astronomers) by many generations of star formation, and where the ingredients are concentrated in the plane of the Galaxy, relatively fewer massive stars appear.

Though the fainter, lightweight stars live for billions of years and the most massive, highly luminous giants for only one thousandth as long, it is the bright stars that initially dominate the regions in which they appear. As soon as they begin to shine, and long before they have settled into their brief, spendthrift lives, energy from the most massive stars in a group begins to destroy and disperse the gas and dust from which they condensed. In doing so they produce some of the most colourful and photogenic celestial displays. In this chapter we present photographs of many such nebulae, each of which has been selected to show some different aspect of this many-sided phenomenon.

## STAR-FORMING REGIONS IN SAGITTARIUS

The M20 nebula, which we saw in the previous chapter, is a particularly beautiful and well-known example of a nebula inhabited by young stars. That image was made to reveal its faint, blue reflective outskirts, and shows where the red emission region fades into the surrounding dust.

Although that coloured boundary marks the limit of the ultraviolet light, there is no physical division here, and tiny dust particles pervade the whole region. Both the emission and reflection nebulae are produced by the small group of hot stars at the centre of the gas cloud, near where three dark lanes seem to meet. To the early searchers of this part of the sky, the dusty patches seemed to divide this bright nebulous region into three parts, hence M20's popular name, Trifid, which simply means cut into three.

Fig. 5.24 was made from three black and white exposures made with the UK Schmidt Telescope. These plates had been specially 'massaged' using a series of unsharp masks to bring out the faintest details, especially the encircling reflection nebula. A more familiar representation of the Trifid nebula is presented in Fig. 6.1. It was made recently from plates taken on the AAT, but I combined the three colour separations without the use of unsharp masks, using instead careful control of contrast to preserve colour balance.

With the lower contrast, the colours are less strident but just as informative, and much more structure can be seen in the dust lanes crossing the emission nebula. The Trifid also contains a well-known and brighter reflection nebula which is clearly shown here. This nebula surrounds a star to the north of the emission region. The star is not hot enough to excite the surrounding hydrogen into its familiar ruddy glow, but it is very luminous, and emits enough light to illuminate a large, partly transparent dust cloud.

When I copied these recent AAT exposures with the same unsharp masking process that was used earlier with the UK Schmidt plates yet more detail emerged, and appears in Fig. 6.2. New features abound, especially in the fragmented dusty patches that seem to separate the red and blue nebulosities. New colours appear as well, and, to me, most interesting of all is the yellow-green patch to the west (right) of where the blue and red nebulae meet. Here, some of the blue reflection nebulosity is partly obscured by foreground dust. Where the dust is thin it absorbs most of the blue light, but some reflection nebulosity remains, fainter, and now a curious greenish colour. This is the first colour picture to show interstellar reddening of a blue nebula.

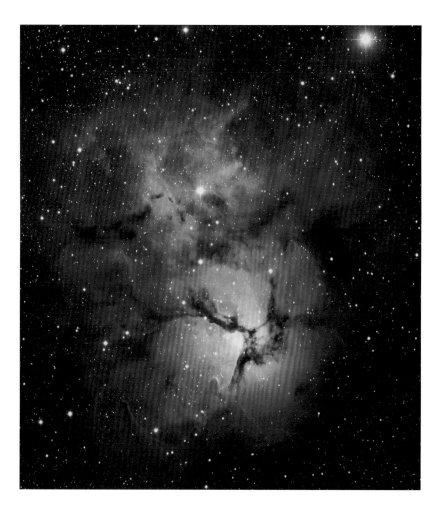

*Fig. 6.1.* A new picture of a famous nebula made recently from black and white AAT plates. M20, the Trifid nebula, surrounds a star-forming region in the southern constellation of Sagittarius. It is a beautiful combination of blue reflection and red emission nebulosity.

At the heart of the Trifid is a compact group of stars that seem to lie near where the dust lanes meet. Within this group, one star is especially massive and it provides much of the energy for both the red emission region and the reflection nebula surrounding it. The radiation from the hottest of the stars is heating and destroying the fragile molecules associated with the dust grains, releasing and dispersing the small particles and their associated gas. Thus the brilliant stars are gradually etching the dark clouds away. This is occurring at the bright rims that we see in Fig. 6.3, which is a more detailed view of the central part of the Trifid nebula.

The brightest rim of all is the curious, blueish 'shoulder' nearest the central stars. This shows that the dust lanes are intimately associated with the stars. As we have seen elsewhere, the fine particles of the dispersed dust grains are effective scatterers of short wavelengths, and towards the centre of Fig. 6.3 the red of the hydrogen is significantly diluted by blue reflection nebulosity.

This new picture of the familiar Trifid nebula reveals a beautiful luminous cloud, which shades from a definite magenta (blue-red) at its centre into a much purer red, and beyond that, quite suddenly back to blue again. This nebulosity is at a distance of about 4500 light years and is about 40 light years across, ten times the distance from the Sun to

*Fig. 6.2.* The plates used to make Fig. 6.1 contain information about the faintest parts of the nebula which can be extracted by special techniques. This shows that the reflection nebulosity completely encircles the emission region and is itself highly structured.

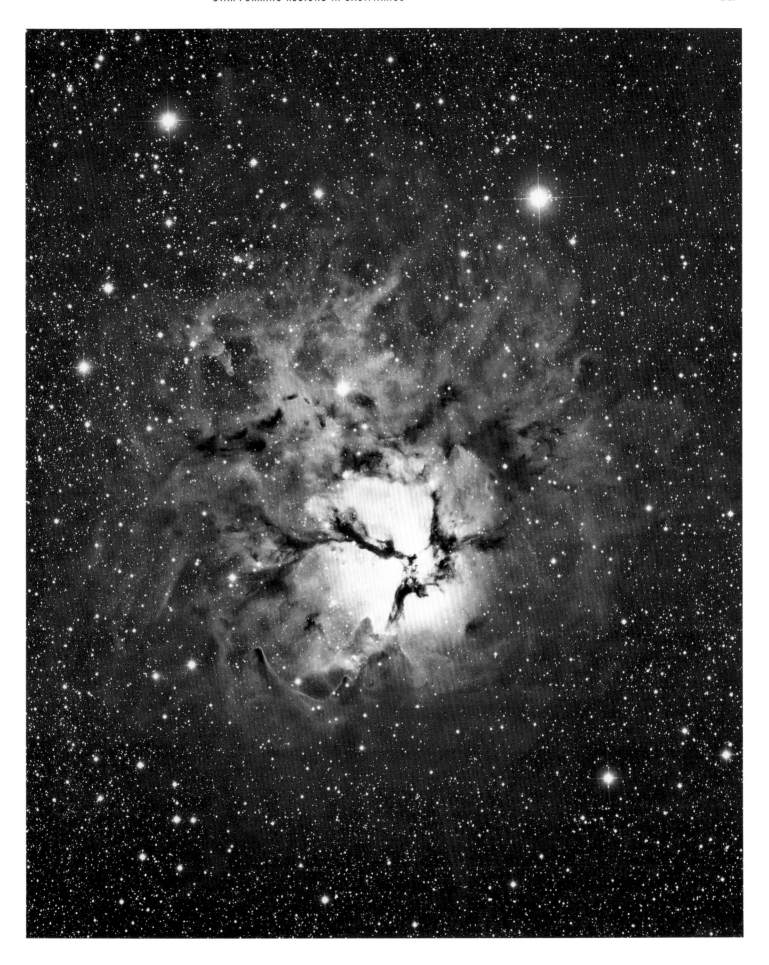

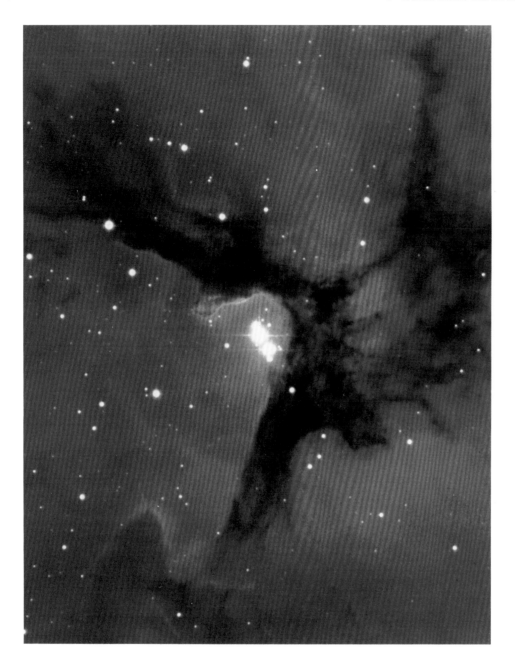

*Fig. 6.3.* At the heart of M20, at the junction of its dust lanes, is a compact group of stars that provide the energy to excite a cloud of hydrogen 40 light years across. The bright rims outlining the dark lanes show that the stars and dust are intimately associated.

the nearest star. The radiation from the handful of bright stars in the Trifid nebula thus dominates a huge volume of space.

Picturing how stars, dust and gas interact in and around star-forming regions is clearly central to their understanding. Although the phenomena illustrated here have been known for many years, and colour pictures of this most beautiful object have been seen since about 1960, this series of new images is an example of the way that complex astrophysical phenomena can be revealed and better understood by means of colour photography, especially if the images are made by colour separation techniques and incorporate advanced image enhancement processes.

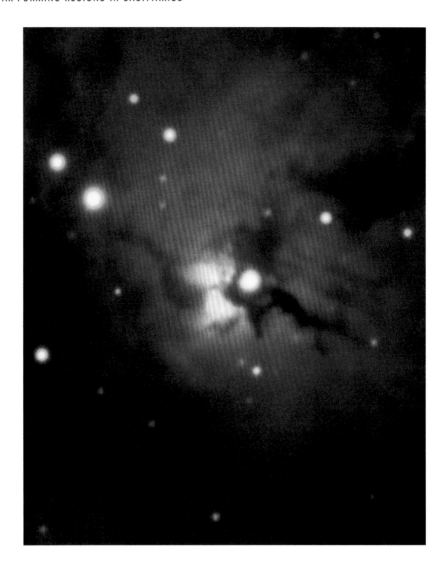

*Fig. 6.4.* This curious bipolar nebula in the brightest part of M8 is the Hourglass nebula, and alongside is the star Herschel 36. Both these objects are among the youngest known, with ages measured in tens of thousands of years.

The colourful nebulosities of the Trifid nebula are only a few degrees from a more extensive nebula that also has a popular name – two, in fact. The enormous Lagoon nebula, M8, contains within it a tiny, very luminous object, known from its shape as the Hourglass nebula. It is seen in Fig. 6.4. Like the Trifid nebula, the Hourglass nebula combines an emission and a reflection nebula, but their various components are not seen separately, probably because the Hourglass nebula is very young. The bright star immediately adjacent to it is Herschel 36, and it is responsible for the nebula. This star is also one of the youngest known. It is probably less than 10 000 years since it emerged from the dense cloud with which it is still intimately involved.

In Fig. 6.5 we see the Lagoon nebula itself (NGC 6523), which is just visible to the unaided eye as a fuzzy patch in the southern constellation of Sagittarius, not far from the conspicuous star clouds seen in Fig. 4.16. The centre of star-forming activity in NGC 6523 is now in the western (right) side, associated with the Hourglass in the brightest part of the nebula. However, the cluster of young stars in the eastern half of the nebula, NGC 6530, testifies to a relatively recent burst of star-forming activity there too.

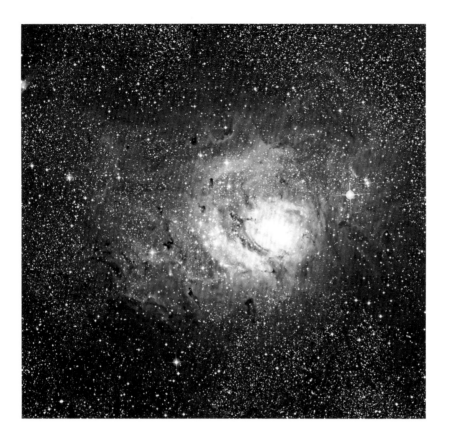

*Fig. 6.5.* The Lagoon nebula (M8) is so named because of the dark striation running across its western (right) third. It contains a cluster of bright stars (NGC 6523), and many dark clouds and bright rims are revealed by unsharp masking.

Though they are young, these stars have been shining brightly much longer than those appearing in the Hourglass region and they have had time to blow the dust around them into a dark swathe dividing the two parts of the nebula. The dusky strip crossing the bright nebula was seen by the early visual astronomers as a dark lagoon and this feature gives the object its popular name. Today's deep pictures show that, far from being a still lagoon, the dust lane is buffeted from within by hot stars, and from without by the still-expanding remnants of ancient supernovae.

M8 and M20 are probably part of the same gassy complex that is seen in Fig. 6.6. One of the advantages of a wide-angle telescope is that it sometimes captures more than one object in the same field. In the richly endowed constellation of Sagittarius this is quite often true, and I made this picture from the same three UK Schmidt plates that were used for Figs. 5.24 and 6.5. The wide-angle view shows the many fainter nebulae in this part of the sky. The shading of star colours across the picture and the variation in the numbers of the stars reveals the patchiness of dust across the field. This picture was made directly from the original plates without any attempt to preserve the brighter details of the more luminous nebulae, so both M8 and M20 lack the visual interest of the previous images.

To the left (east) of the Lagoon nebula, and seemingly linked to it, is a complex group of fainter nebulae. When I copied the black and white original images of this region at a higher contrast to show fainter details before combining them into a colour image, this part of Sagittarius suddenly burst into a profusion of colourful nebulae.

Fig. 6.7 shows one of the most beautiful patches of nebulosity anywhere in the sky. In the foreground we see both emission and reflection nebulae set against the clouds of yellow stars that swarm around the nucleus of the Galaxy. These are NGC 6559 and IC 1274–5. The dark spaces between the bright nebulae are the unlit portions of one of the numerous molecular clouds that hide so much of the Galaxy in this direction. Where new stars have appeared, the dark molecular clouds are lit from within, sometimes revealing the dust as sinuous dusky lanes crossing the brightest parts of the luminous backdrop. Soft blue reflection nebulae mark the regions where light from the bright stars is eating its way into the molecular clouds. This interface is marked by the change in colour, from blue to red, as in M20.

## THE ORION REGION

The nearest and best-studied star-forming region is that in Orion. At a distance of about 1500 light years, and much closer than the nebulae in Sagittarius, Orion is close enough and compact enough for us to see both the finer details and the large-scale structure. The widest view of all in our series of portraits is seen in my picture of Barnard's Loop, which appears in Fig. 7.8. It was made with a conventional plate camera, but a more detailed look at the principal nebulae is shown in the UK Schmidt photograph, Fig. 6.8, which contains both the Horsehead nebula and, at the bottom of the picture, the famous Orion nebula, M42 itself.

The brightest stars, to the right of the Horsehead and at the top of the picture, are the easternmost pair of the line of three similarly bright stars that define Orion's Belt. This line of stars makes the constellation so distinctive, and is the most obvious feature of the constellation in the star trails above Coonabarabran that we saw earlier.

The region of Orion between the Horsehead nebula and M42 is replete with the dusty ingredients of star formation, some of which feebly reflect the light of the many highly luminous stars and nebulae. The numerous faint wisps in Fig. 6.8 are evidence of this, and deeper images, massaged in the darkroom from long exposures on the UK Schmidt Telescope (Fig. 6.9) show that even the dark cloud from which the Horsehead seems to emerge is faintly luminous. Throughout this region, radio astronomers have found extensive molecular clouds, most of which are invisible at optical wavelengths. Though springing from clouds that seem mostly dark and quiescent, in some parts of Orion the business of star formation is very active indeed, and this leads to much more dramatic images, none more so than the Orion nebula itself.

At the brightly-lit core of this nebula, as in the more distant Trifid nebula, a few extremely hot stars are responsible for most of the action. In the Orion nebula they appear as a compact but distinctive group known as the Trapezium cluster, just resolvable as individuals in binoculars. With a large telescope, such as the AAT, the distinctive shape of the formation can be seen. Even with very short exposures, such as the one-minute twilight 'snapshot' that I made at the AAT's $f/8$ focus, the background nebulosity tends to dominate the stars, and unsharp masking had to be used to retain the images of the fainter members to produce Fig. 6.10.

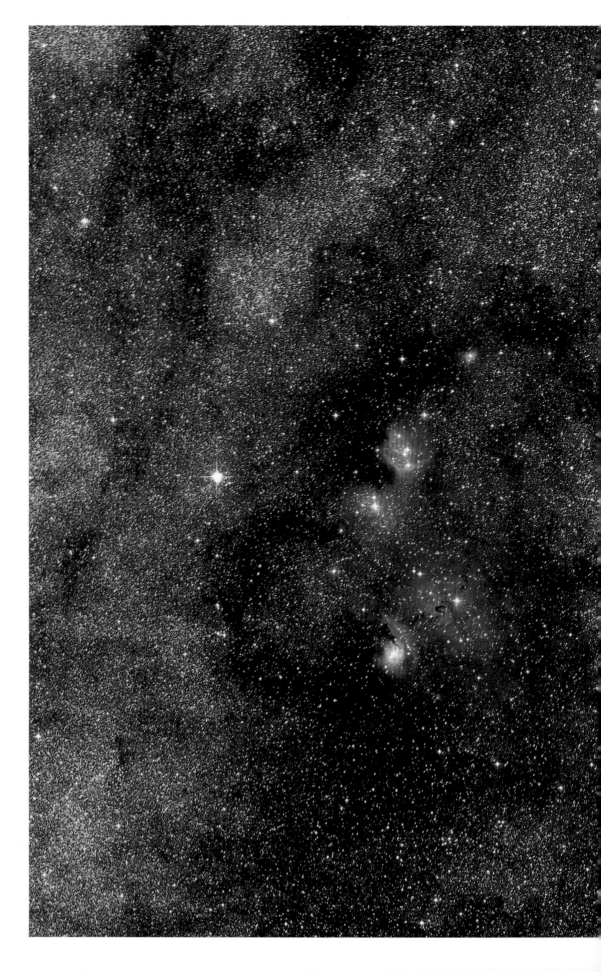

*Fig*. 6.6. A wide-angle view of the beautiful nebulosities in Sagittarius shows the NGC 6559 region (left), and the Trifid (top) and Lagoon nebulae as they appear when the original plates are combined without image enhancement. This same set of plates was used to make Figs. 5.24, 6.5 and 6.7.

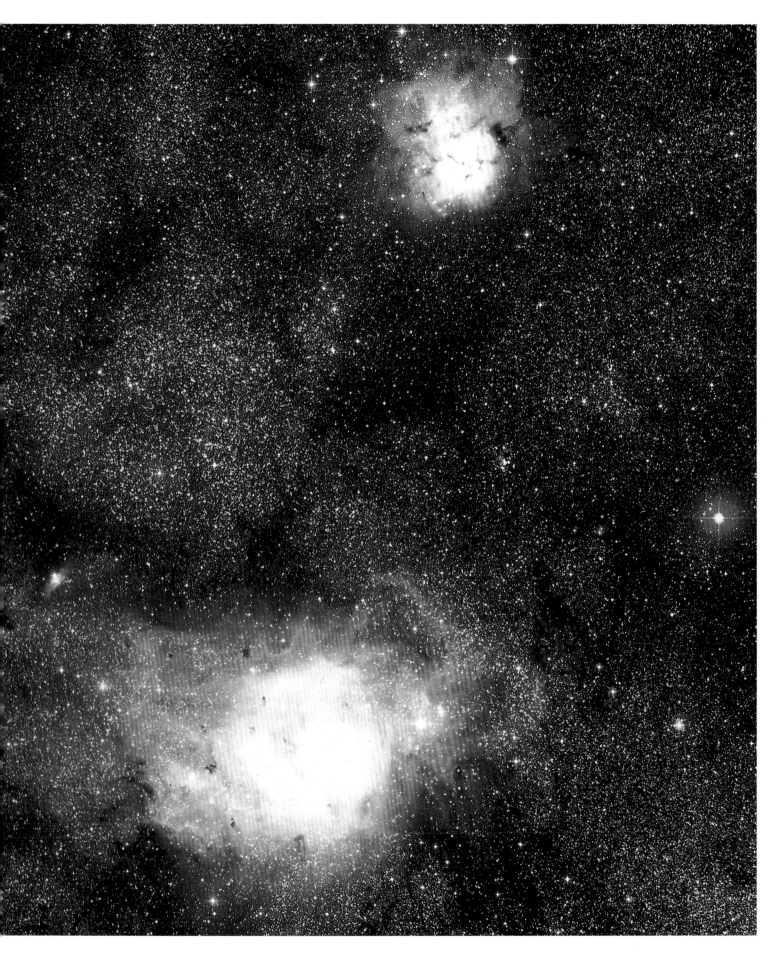

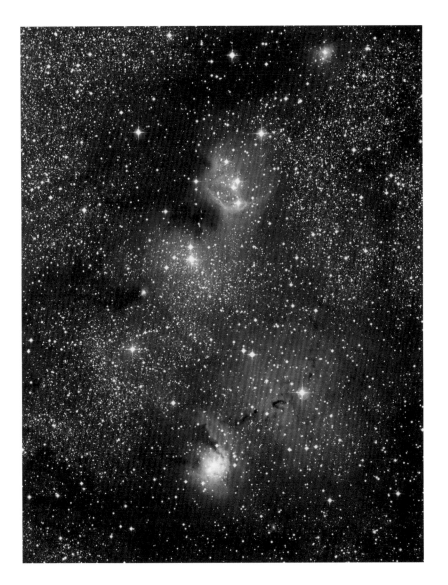

*Fig.* 6.7. Faint nebulosities of NGC 6559 and IC 1274–5 are almost certainly part of the same gas cloud as the nearby Lagoon and Trifid nebulae. Here, illuminated by a few bright stars and on a smaller scale, are seen several reflection nebulae surrounding red emission regions.

The Trapezium stars are seen in their wider context in Fig. 6.11, a colour picture made from three even shorter exposures that I made in twilight at the *f*/3.3 prime focus of the AAT. (Dark time on a large telescope is too valuable to be used for short exposures.) The Trapezium group can now be seen as blue stars against the multicoloured nebula behind them. They illuminate one of the few emission nebulae that can be seen with the unaided eye. In the brightest parts of the Orion nebula, those parts of the molecular cloud closest to the searing radiation from the Trapezium stars, oxygen as well as hydrogen is excited into fluorescence, and it emits a green light of sufficient intensity to be on the threshold of colour vision. With a telescope, the keen-sighted tell me they can see this part of the Orion nebula glowing with a pale greenish hue, but it always appears colourless to me, even with the AAT.

The equally intense red light from excited hydrogen is not seen because it is a part of the spectrum to which the human eye is relatively insensitive. No such limitations affect

*Fig.* 6.8. The full 6.5 × 6.5 degree field of the UK Schmidt Telescope includes all the most spectacular nebulae in Orion as well as two of the three bright stars that are in Orion's belt. Without unsharp masking, the Orion nebula (bottom) appears as a colourless blob, but the fainter wisps of surrounding dust and diffuse red emission nebulosity fill the field.

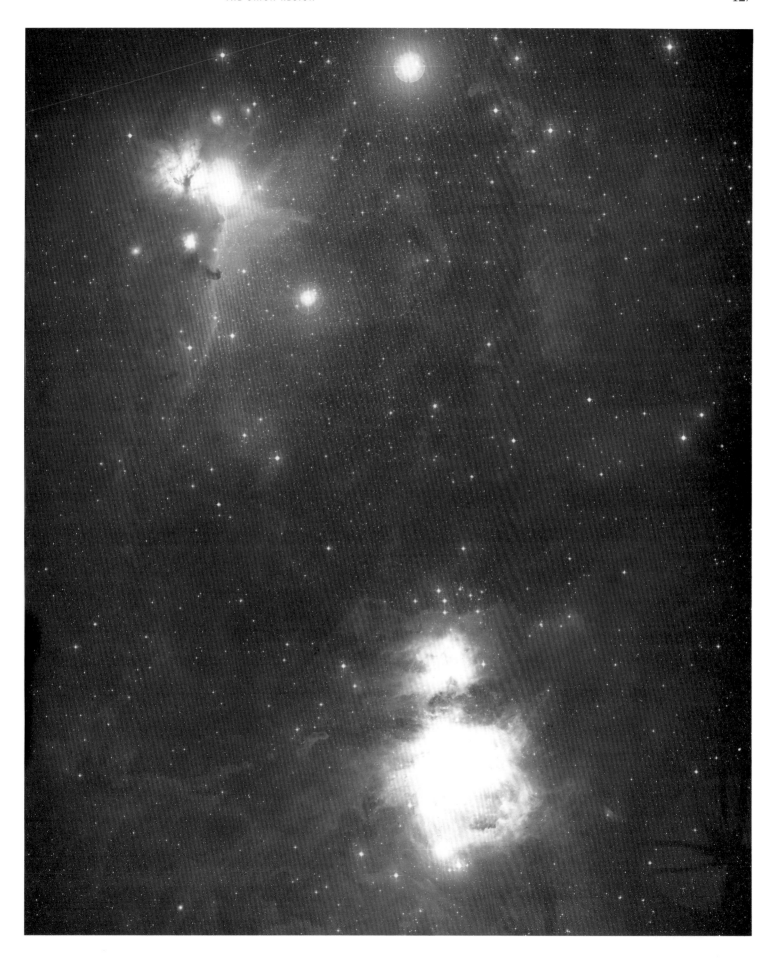

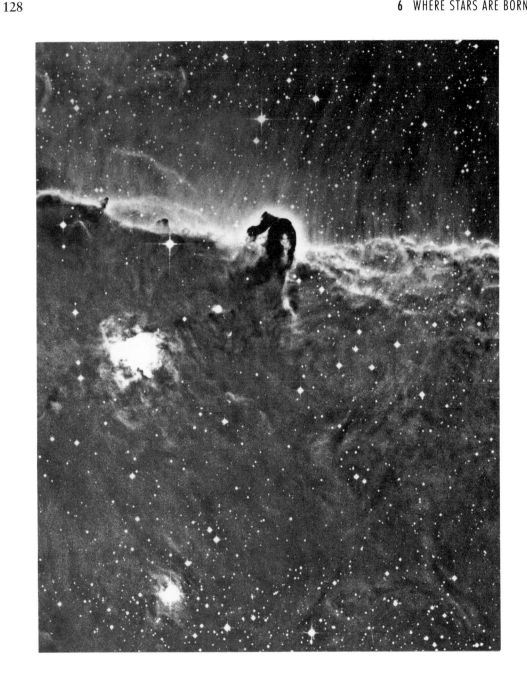

*Fig. 6.9.* On the 'dark' side of the Horsehead nebula there is a large region of faint nebulosity that is best seen in black and white. As well as the bright strip of the emission nebula IC 434 it contains two embedded reflection nebula, NGC 2023 (near the horsehead) and IC 435 at the bottom of the picture. This picture was extracted from the red-light plate used for Fig. 6.8.

the photographic plates used for these pictures. They can garner faint light over long periods and be combined to make a picture that represents the object much as the eye would see it if the object itself were much brighter, and its components of higher contrast.

The mixture of green and red emission dominating the central regions is responsible for the yellowish hue seen in these photographs, but this is diluted by scattered starlight and relatively weak lines in the blue part of the spectrum. Unlike *subtractive* colour mixtures made with pigments or dye-stuffs, where red and green would produce a darker shade of brown, here we are mixing light *additively,* and green and red together produce

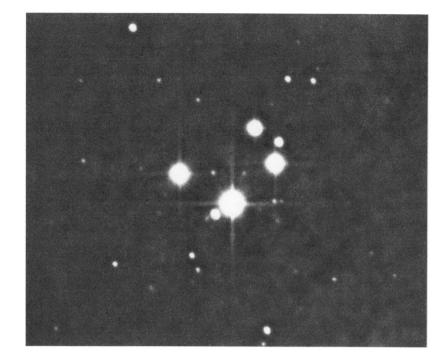

*Fig. 6.10.* A high resolution picture of the Trapezium group at the centre of Orion's brightest nebulosity was made on ordinary black and white film at the AAT's *f*/8 focus. It has been copied by unsharp masking to suppress the bright nebulosity and emphasise the fainter stars.

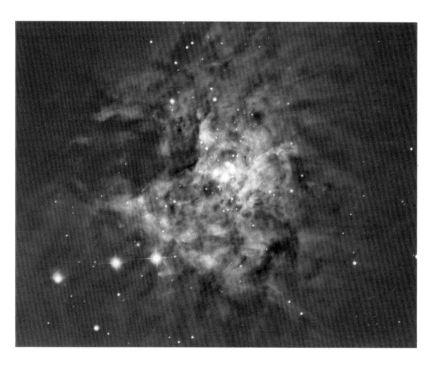

*Fig. 6.11.* Three 1-minute exposures were combined to reveal the colour of the inner Orion nebula. At its heart are the Trapezium stars. These are the most obvious members of a young but substantial cluster of stars that are mostly hidden in the bright background.

a brighter shade, in this case yellow. To add to this colour confusion, with ordinary colour film this region normally appears red.

Unlike the eye, or the black and white plates used here, conventional colour film is relatively insensitive in that narrow part of the spectrum where the green oxygen lines occur, but not in the similarly small part where the red light of hydrogen appears, so the

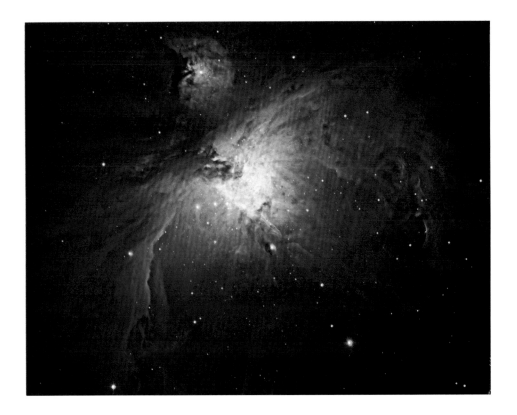

*Fig. 6.12.* A series of five-minute exposures shows the fainter parts of the inner regions of M42, the Orion nebula, and especially emphasises the multi-coloured, curved rim of the cavity created by the Trapezium stars.

red predominates and gives an inaccurate colour rendering on conventional colour film. This should not be surprising; colour films are intended for recording the broad band colours of everyday life, not the tenuous glowing gas between the stars. The three-plate system of colour photography that I use is covered in more detail in the references in the Bibliography and it entirely overcomes these problems. Though that process is ideal for astronomy, it is highly inconvenient for 'everyday' picture-taking.

The complex structures and filaments seen in Fig. 6.11 and a wider view, Fig. 6.12, made from three 5-minute AAT exposures, have been emphasised by enhancing the image by an unsharp masking process applied to the black and white originals before the colour picture was made. It is clear from the illusion of movement that the enhanced detail creates in these pictures that the Trapezium stars are the source of an energetic outflow in the nebula. Such outflows are a normal consequence of the birth of massive stars, and the radiation and stellar winds flowing from them will eventually disperse the surrounding gas and dust that spawned the stars.

Already, after a lifetime of much less than a million years, the Trapezium stars have blown a hollow in the dense nebula. The interface between the rapidly dispersing nebula and the resistant bulk of the material can be seen as the sharp-edged curve running across the upper left of Fig. 6.12. The colour change indicates that we are looking into this stellar nursery from above one edge of a bowl-like hollow, rather like peering into the parabolic reflector of an automobile headlamp, with the Trapezium stars as the glowing filament at its focus. On one side of the rim, the gas is directly illuminated by the Trapezium stars, on the

*Fig. 6.13.* The Greater Orion nebula. Made from long exposure Schmidt plates, this picture reveals the red bubble of hydrogen blown out from the core of the nebula by radiation from the Trapezium stars.

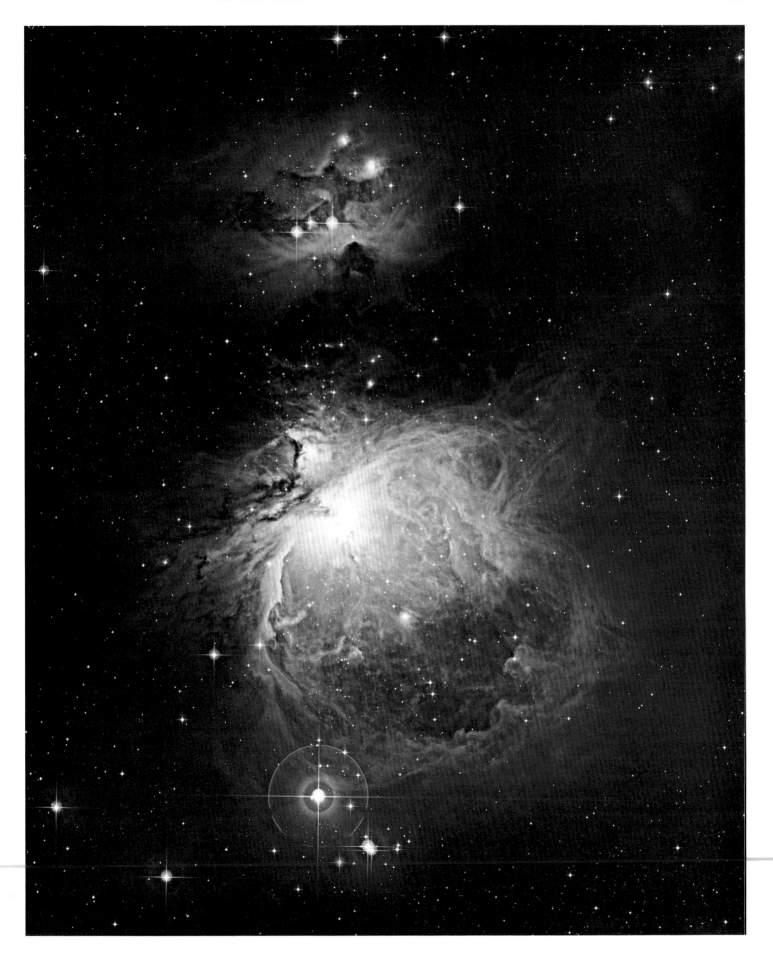

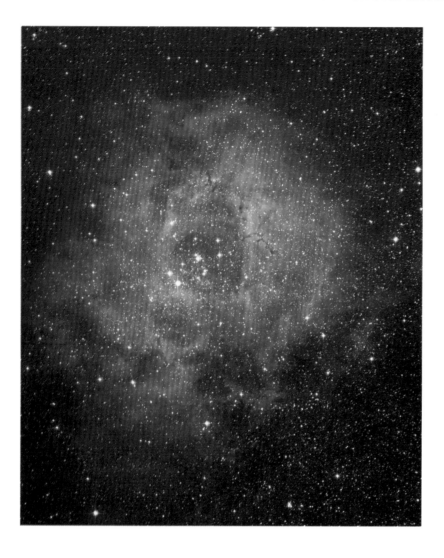

*Fig. 6.14.* The aptly-named Rosette nebula in Monoceros surrounds a cluster of brilliant, massive young stars.

other it is not, which accounts for the clearly defined edge and strong colour difference.

The brightest part of the nebula is also the densest, and for the moment it is mostly resisting erosion by radiation from the stars. However, the stars must have formed in what seems to us the south-western outskirts of the molecular cloud and in that direction (lower right in Figs. 6.11 and 6.12) most of the material has been blown clear. Largely unimpeded, the intense ultraviolet light from the stars now illuminates an expanding sphere of hydrogen. The Trapezium stars at the heart of the nebula are thus responsible for the astonishing faint red wispy nebula seen on its outskirts in Fig. 6.13, probably the most intriguing picture I have ever produced by the unsharp masking process.

It should be explained that Fig. 6.13 was made from the same plates that were used for Fig. 6.8 (and Figs. 5.3 and 5.17). The Orion nebula is so bright that it appears grossly overexposed on long-exposure photographs. Fortunately, because this picture was made from three black and white plates, overexposure does not necessarily lead to loss of infor-

*Fig. 6.15.* NGC 2244, the star cluster in the Rosette nebula. All the blue stars are members of the cluster that excites the nebula but the brighter, yellow star is in the foreground and not part of the group.

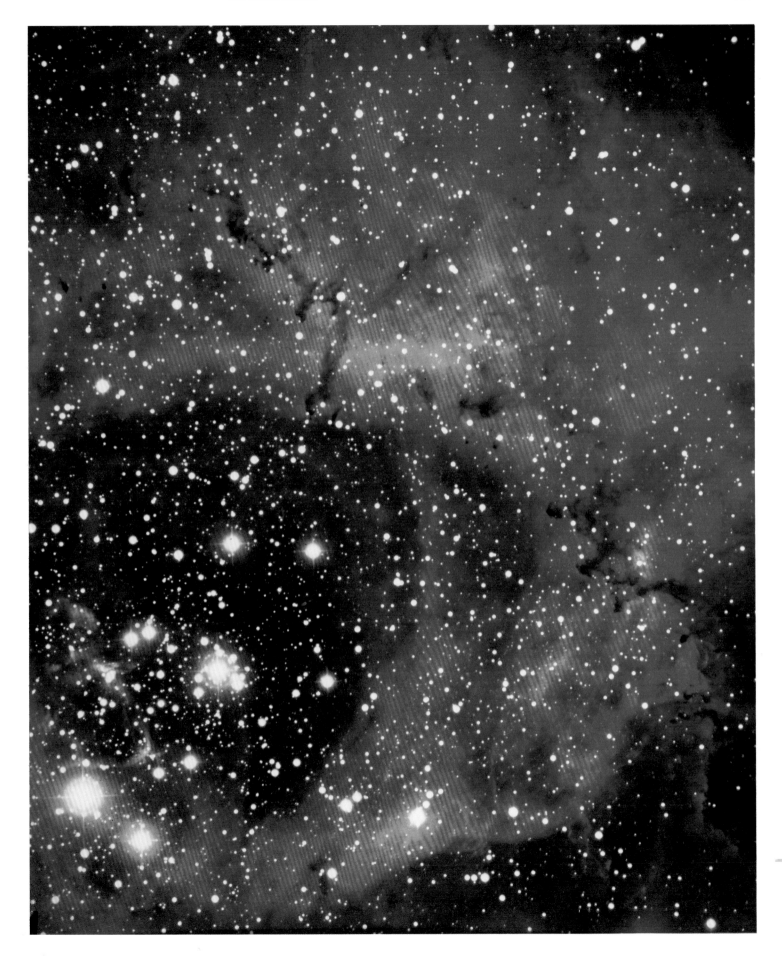

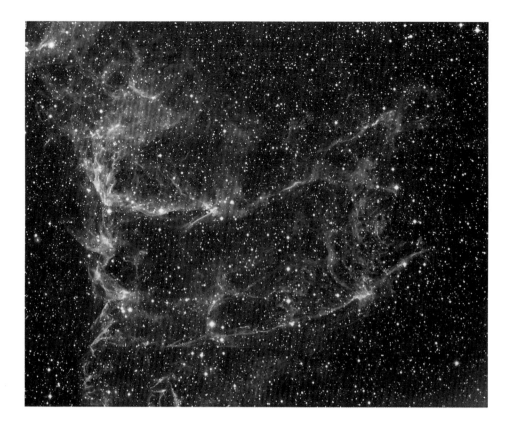

*Fig. 6.16.* Part of the Veil nebula, a middle-aged supernova remnant.

mation, as it would if colour film had been used, so the hidden images can be extracted from the plates with unsharp masking before they are combined into a colour picture. As a result, information apparently missing in the rather uninteresting blob of the Orion nebula in the lower half of Fig. 6.8 can be extracted to reveal the beautifully detailed structures seen in Fig. 6.13.

## A GALLERY OF STAR-FORMING REGIONS

The vigorous dispersal of the material from which they formed is a common feature of the nebulae associated with the birth of stars. It is clearly evident in the aptly named Rosette nebula, seen in its entirety in Fig. 6.14. A more intimate view, Fig. 6.15, shows the colours of the hot stars that comprise the central cluster, NGC 2244, and the enormous cavity they have created. The hollow is about 12 light years across and the luminous walls are expanding at about 20 kilometres per second, but, unlike the Orion nebula, the stellar winds from the stars seem to have met little resistance in any direction, so the nebula is remarkably symmetrical from our point of view. Soon, within the next million years or so, these energetic stars will blow the nebula away completely, probably aided by one or more of them exploding as supernovae, leaving a brilliant but depleted young open cluster ornamenting the constellation of Monoceros. Of course, supernovae can also produce distinctive nebulae, which often appear as fragments of expanding shells like that shown in Fig. 6.16, part of the Veil nebula.

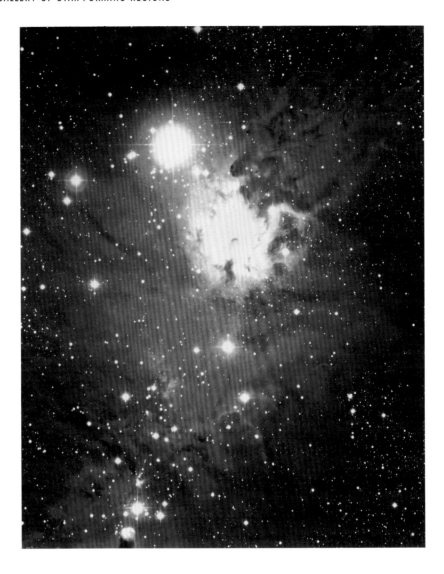

*Fig. 6.17.* The faint nebulosity and Herbig–Haro objects around S Monocerotis, the brightest star in the picture. The tip of the Cone nebula is visible at the lower edge of the photograph.

The interplay of light and dark in these nebulae provides ample evidence that dust is intimately associated with star formation. This is also very obvious in Fig. 6.17, a nebula without a name, also in Monoceros. In the upper centre of the picture, to the left of the brilliant star S Monocerotis, is a blueish reflection nebula, and to its right a most curious striated emission nebula, which has always reminded me of a shiny red fox fur. It is part of a larger star-forming region 2000 light years away and covering several degrees of the sky. Also here, evident as curious low-contrast wisps and loops, are some of the largest Herbig–Haro objects known. These low-contrast features are often associated with outflows from young stars and are described in more detail in the next chapter. In the lower part of Fig. 6.17 the northern tip of the Cone nebula appears. It is seen more clearly in Fig. 6.18. The 'cone' itself is similar to the Horsehead nebula, a dusty protruberance illuminated by a bright star and illustrated in Figs. 5.3–5.6. It is not at all unexpected to find Herbig–Haro objects here.

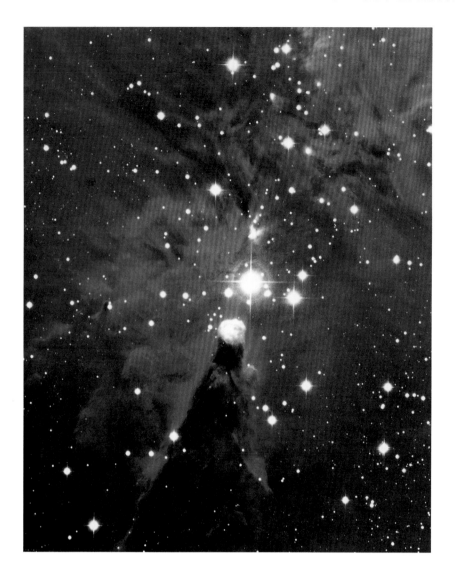

*Fig. 6.18.* The Cone nebula surrounded by the scattered bright blue stars of the open cluster NGC 2264.

As in the Horsehead, the vivid red back-illumination is provided by the glowing surface of a molecular cloud illuminated by several very hot stars. This image is one of the few made at optical wavelengths to reveal the presence of newly formed but still hidden stars, not yet free of their dust cocoon. In truth, they were first detected by my colleague David Allen at infrared wavelengths, but they reveal their presence in these pictures by the yellow colour they impart to the dust that embraces them.

Light from these young stars is strongly scattered, partly absorbed and then reflected by the dust. This tortuous path gives a distinctly yellow shade to the light emerging feebly from chinks in the obscuring veils. Soon, these stars too will blow away their dusty fronds and become as visible as those in the foreground, which were formed in a similar dark place relatively recently, about 3 million years ago. For the moment they remain the hidden precursors of the next generation of stars.

The extent of the dust cloud is seen on Fig. 6.19 by virtue of the stars it hides. This is a new UK Schmidt picture of the region around the Cone nebula, and problems with

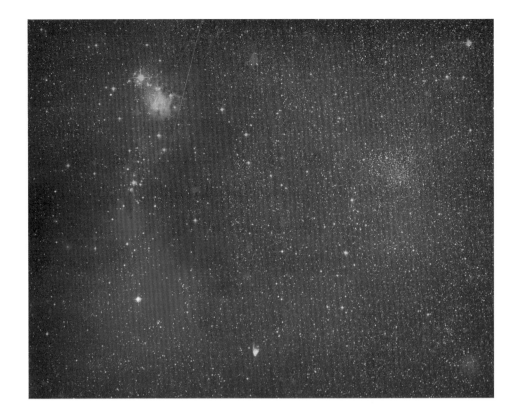

*Fig. 6.19.* A wide-angle view of the Cone nebula region also shows Hubble's variable nebula (NGC 2261) as the small white blob at lower centre and the old open cluster Trumpler 5 at west (right) of the frame. (See also Fig. 4.20.)

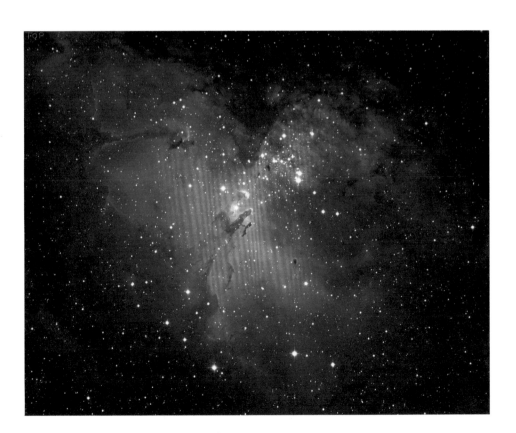

*Fig. 6.20.* The dusty nebulosity of M16 is associated with the open cluster NGC 6611.

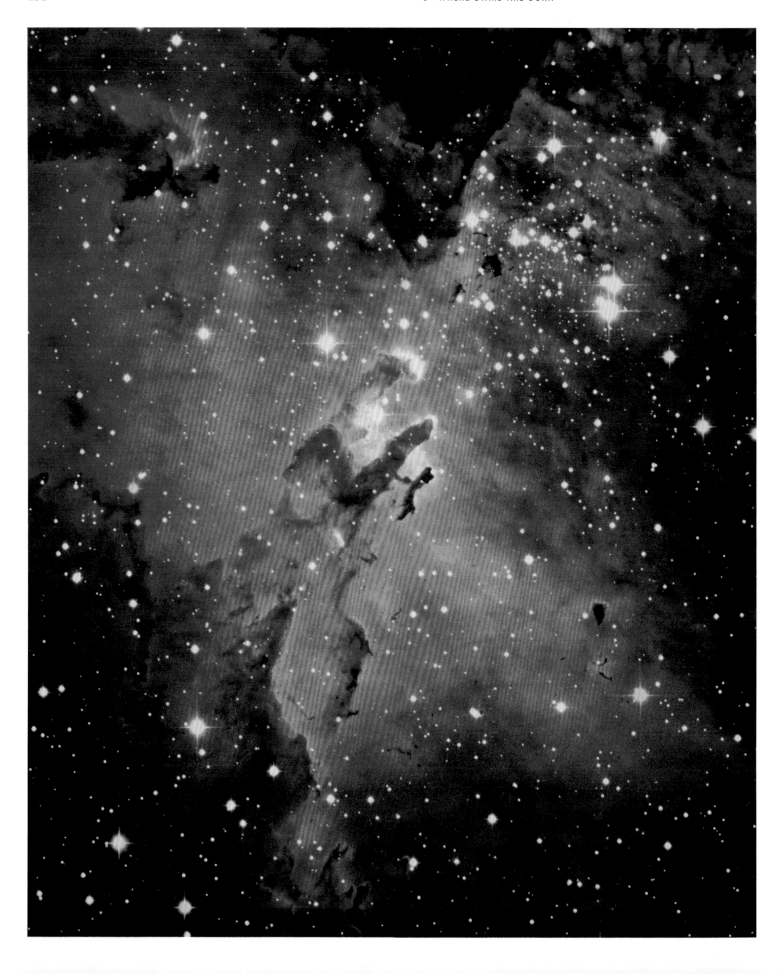

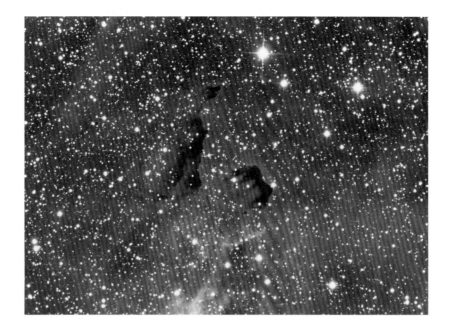

*Fig. 6.22.* A large Bok globule seen against the nebulosity of NGC 3576 in Carina.

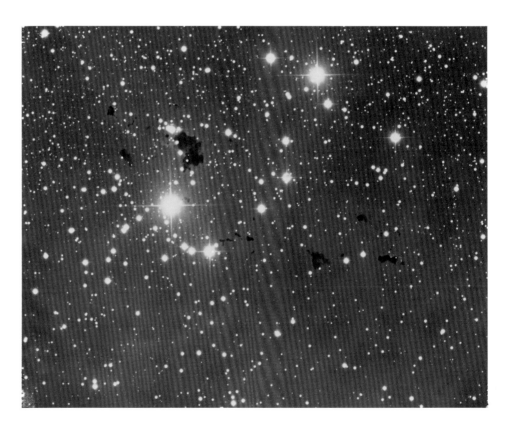

*Fig. 6.23.* A beautifully defined group of Bok globules seen against the large, faint southern emission nebula IC 2944.

*Fig. 6.21.* The distinctive 'elephant trunk' dust lanes in M16, the Eagle nebula, are elongated by the action of stellar winds from the massive stars in NGC 6611. A few dark Bok globules are scattered across the face of the nebula.

the exposure of one of the plates have prevented me from achieving perfect colour balance, but it is included here because this is the first colour image of NGC 2261, Hubble's variable nebula, seen as the small, white cone-like image at the lower centre of the picture. Also here is the old open cluster of stars Trumpler 5, at the right (west) of the pho-

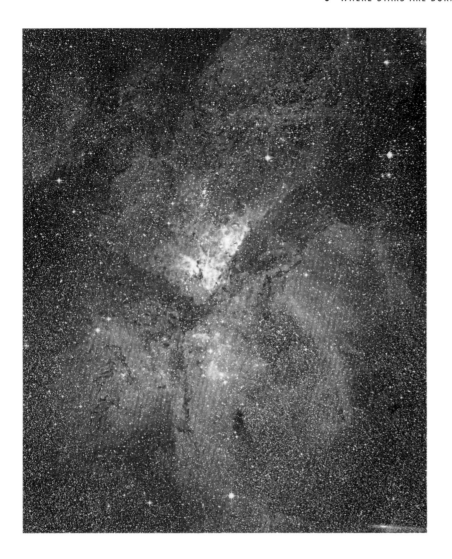

*Fig. 6.24.* The Carina nebula (NGC 3372) in the southern Milky Way is one of the largest and most spectacular star-forming regions in the Galaxy. On deep plates the nebulosity can be traced out over many square degrees of sky.

tograph. Its identity as a cluster is more obvious here than it was in the AAT picture used as Fig. 4.20.

Other cone-like dusty nebulae under attack by starlight are to be found, and among the most conspicuous are the well-known 'elephant trunks' in Messier 16, the Eagle nebula (Fig. 6.20). These structures have bright rims and seem to point themselves at the young cluster of stars, NGC 6611, in the upper right of the photograph, the source of the destructive radiation. This nebula is also famous for its 'Bok globules', better seen in Fig. 6.21 as small, usually isolated clouds of dust, which often have bright rims when associated with bright nebulae. Many such are known, but Fig. 6.22 shows a magnificent example, seen against the curious loops of NGC 3576, which has not been noted previously. Another fine collection of Bok globules is seen against a large emission nebula, IC 2944, in Carina (Fig. 6.23).

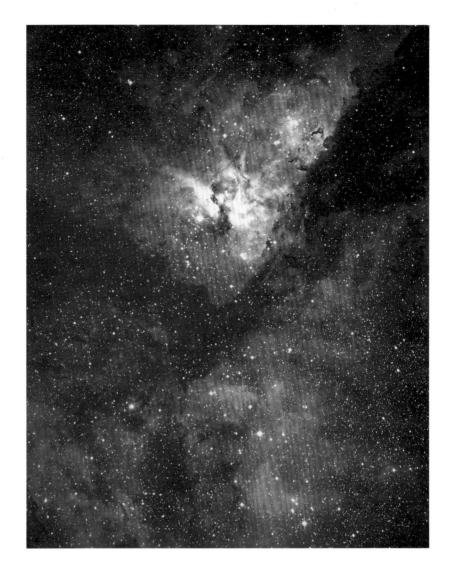

*Fig. 6.25.* The dark lane that appears to divide the Carina nebula is just visible in binoculars against the backdrop of stars. This AAT picture reveals the numerous dust clouds and stellar groups caught up in the nebulosity.

## THE ETA CARINAE REGION

The constellation of Carina contains one of the most spectacular regions containing young stars in the sky. Unfortunately for most of the readers of this book, it is only visible from southern latitudes. The most prominent of the many nebulosities here is the Carina nebula, NGC 3372. It is between 8000 and 9000 light years distant, yet it is visible to the naked eye, mainly because of the large number of very luminous stars it contains. The fainter tendrils of the nebula are seen in Fig. 6.24; as well as the nebula, this wide-angle UK Schmidt image reveals the enormous number of stars in this direction. An even wider view of the southern sky between Crux and Carina appeared in the previous chapter.

This region is famous not only for its nebulae, remarkable though many of them are. It also contains some of the most luminous stars in the galaxy and they are concentrated in the brightest emission region in the constellation, the spectacular Carina nebula. The brightest parts of this nebula rival the Orion nebula in interest and would have

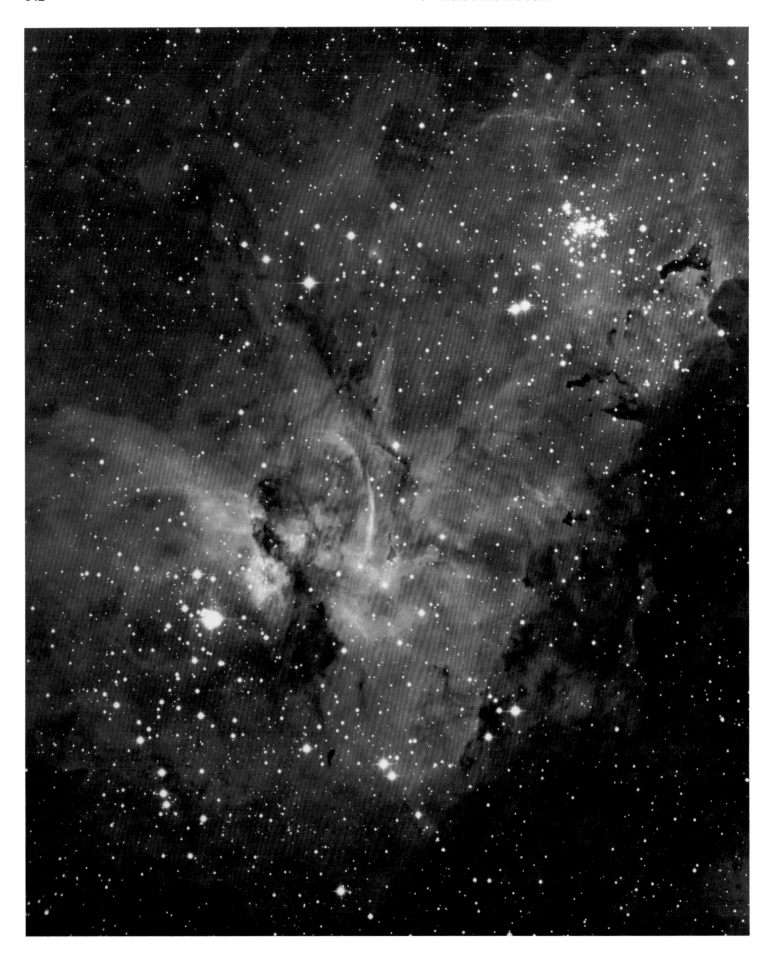

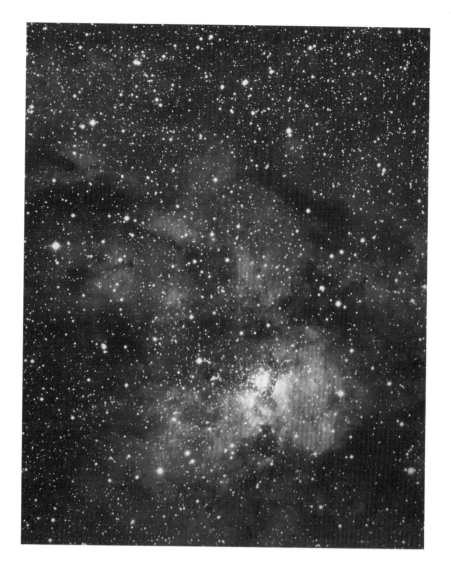

*Fig. 6.27.* The open cluster NGC 3603 and its nebula are almost hidden by Galactic dust. This young, compact cluster contains an unusual number of very hot stars.

*Fig. 6.26.* The brightest stars in this part of the Carina nebula are some of the most luminous known in the Galaxy, some brighter than a million Suns. While most are seemingly isolated, a few are associated with the cluster Trumpler 14 (upper right).

appeared in many early astronomers' sketch books if Carina had been visible from the northern hemisphere. Like the Orion nebula, that in Carina is full of detail on all kinds of scales, so wide-angle images (Fig. 6.25) and more detailed views of the inner part of the nebula (Fig. 6.26) show ever finer structures.

These photographs show as a tiny, curiously misshapen bright nebula one of the most remarkable stars ever seen, even more remarkable for the fact that almost 160 years ago it was briefly the second brightest star in the sky, only outshone by Sirius. This star and its nebula were the subject of much speculation by Sir John Herschel, who observed it from South Africa in the 1830s, when it was near its brightest. It is one of the most extreme examples of a massive star shedding a substantial fraction of its substance; mass-loss stars, and especially the remarkable Eta Carinae, are examined further in Chapter 7, and Herschel's drawings of the star and its nebula are seen in Chapter 5.

The brilliant stars in the Carina nebula seem to be scattered across a considerable volume of space, and apart from a few sparse clusters there are no obvious concentra-

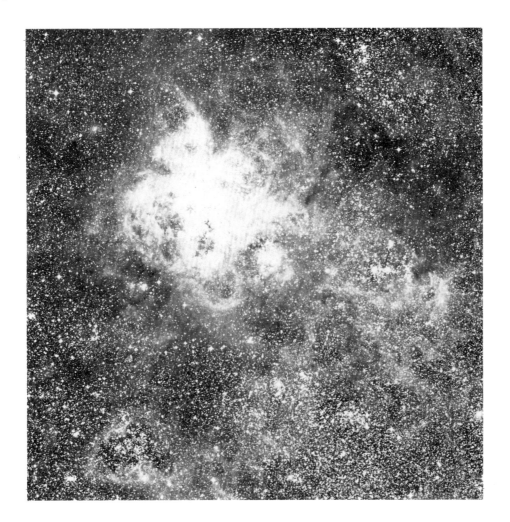

*Fig. 6.28.* The region around the Tarantula nebula in the LMC is rich in clusters of massive stars and traces of the nebulae from which they form. Some dusty patches stand out against the background as darker and yellower than their surroundings.

tions. However, in the same constellation there is another remarkable region where a huge number of very hot stars are concentrated into a very tiny volume. Unfortunately, this compact cluster, NGC 3603, illustrated in Fig. 6.27, is heavily obscured, so is less well studied than might otherwise have been the case.

Within this compact cluster are several Wolf–Rayet stars, some of the hottest and most luminous types of star known. They must have formed there recently, because their extreme properties ensure a short life. No other such group is known in our own Galaxy and if NGC 3603 could be seen without obscuration it would be spectacular indeed. Because such concentrations of stars are so luminous they can be seen through thick layers of dust and over vast distances, even in galaxies beyond the Milky Way, such as the LMC.

## STAR FORMATION IN THE LARGE MAGELLANIC CLOUD

In the LMC is one of the biggest and most productive star-forming regions known anywhere. It is centred on the eastern part of the galaxy and is seen as the brightest nebu-

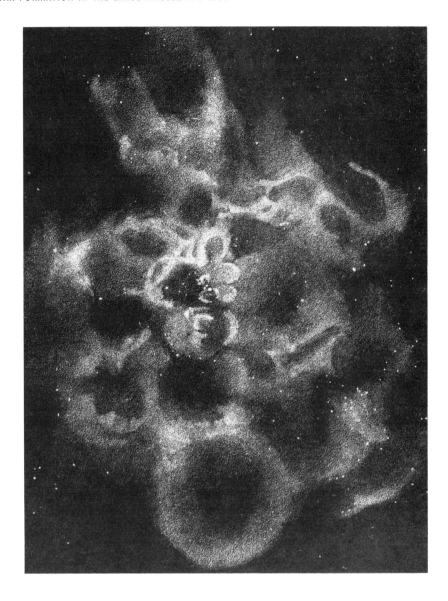

*Fig. 6.29.* The colours of the nebula were not noted by Le Sueur as he sketched the brighter strands of glowing gas with the Great Melbourne Telescope in 1870. It is clear from this representation how the Tarantula nebula acquired its name.

losity to the left of Fig. 8.4. More detail is seen in Fig. 6.28, which reveals the huge number of bright stars that surround it. This rich constellation is 170 000 light years distant, so all the stars visible in the photograph are much more luminous than the Sun. Sun-like stars are certain to be there, but they are too faint to be seen on this wide-angle picture.

Dominating the top (north) of the photograph is the Tarantula nebula, a region where star formation is at its most vigorous. Its name derives from its intriguing visual appearance, well represented by Fig. 6.29, derived from a drawing made by M. Le Sueur, who had a short but productive career as an observer on the Great Melbourne Telescope. It was Le Sueur who found that polishing the speculum metal mirrors was an art rather than a science, and he was less successful with the mirrors than he was with his pencil. The fainter parts of the outer Tarantula are recorded more clearly photographically in

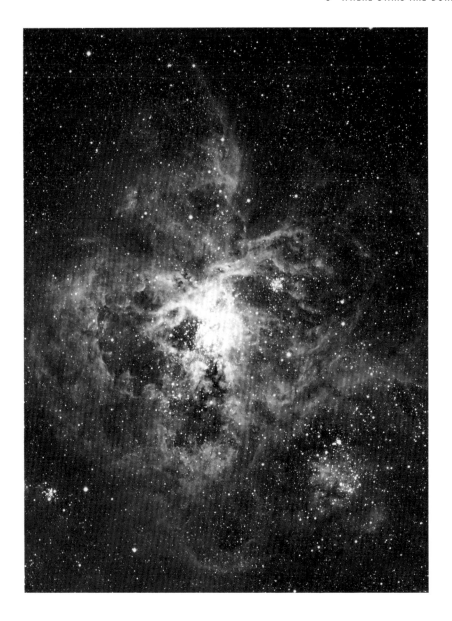

*Fig. 6.30.* This AAT photograph shows nebulosity around the Tarantula nebula much fainter than can be seen by eye, but without special copying the brighter parts shown in Le Sueur's drawing are overexposed.

Fig. 6.30, but even on this high resolution AAT picture the central nebulosities are lost, their brilliance too much even for the most diligent unsharp masking.

At the centre of this vigorous activity is the compact object 30 Doradus. Its name suggests that it is a single star, and it was long thought to be so, since even the most powerful telescopes could not resolve it. A very short, high resolution AAT image (Fig. 6.31) reveals why; at the core of the nebula is a knot of very bright stars, their images blending one into the other. If the powerhouse of the Tarantula nebula were a single star, its apparent brightness would suggest that it was a most extreme specimen, with a mass of perhaps a thousand times that of the Sun. Such stars are unlikely to be stable enough to survive and another explanation had to be found.

Special image recording and analysis techniques designed to overcome the blurring of the Earth's atmosphere have finally resolved the central 'star' in 30 Doradus into sev-

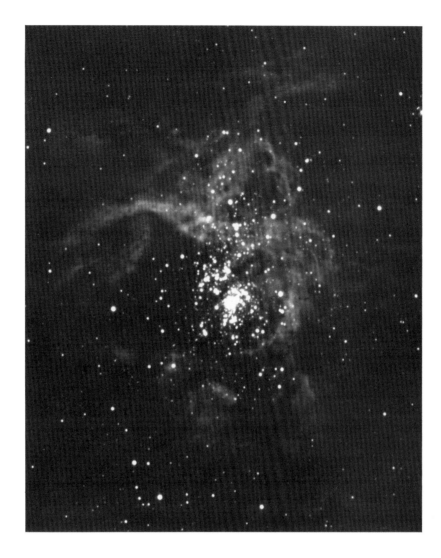

*Fig. 6.31.* A very short exposure shows the cluster of stars at the core of the Tarantula nebula. The largest, unresolved group was once thought to be a single, uniquely massive star, 30 Doradus, but it is now known to be a compact cluster of many very luminous stars.

eral components, an observation confirmed by the Hubble Space Telescope. Many of these stars are extreme in their own way, but not as implausible as single, super-massive 30 Doradus was. Some of them are Wolf–Rayet types like those in NGC 3603 (Fig. 6.27) and the whole cluster is now believed to very similar to the NGC 3603 group, but much richer in very hot stars and 20 times more distant.

The nebulosity in a region with so many luminous stars is buffeted about by supernova explosions and by the intense winds blasting from their surfaces. Sometimes, the stars blow away the nebulosity to form cavities relatively free from nebulosity. The bright rims of these cavities are the spindly legs of the 'Tarantula', seen so well in the Great Melbourne Telescope drawing. The photographs here show that they are widespread throughout the region, and are common wherever relatively stationary gas and dust is affected by energetic winds from massive stars. Stellar outflows are the subject of the next chapter.

# 7 MASS LOSS AND THE LIFE AND DEATH OF STARS

## INTRODUCTION

Why should anyone want to study the stars? We all know they are there, enigmatic points of light, scattered across the night sky, twinkling, romantic. But why should we wish to know more about them?

There are at least two reasons. First of all, human beings have an insatiable curiosity, a need to know and understand, that marks them out from other creatures. That is how we know, after centuries of patient investigation, that the stars are distant Suns. This same curiosity and search for order in the natural world has evolved into the systematic quest for knowledge that we call science. And secondly, scientists have discovered something quite fascinating about the stars. They have found that we are creations of those glittering points of light. Without them we would not exist, nor would the planet on which we spend our lives.

In the beginning there was hydrogen and helium and very little else. These elements were created in the Big Bang along with time, space, everything. At all but the lowest temperatures hydrogen and helium are gases and are, well, uninteresting. Planets, plants or people cannot be made from lightweight elements such as these, but stars can.

In the previous chapters we caught glimpses of the stuff from which stars were made and saw indirect evidence of their birth. But the stars exist, and we will now follow their life and death, and later see how they congregate together in vast numbers in galaxies. These entities, stars and galaxies, seem to be interdependent. One cannot exist without the other, and we could not exist without either, so a study of these objects is, in the end, a part of the quest for the origin of life itself.

Initially, stars are local concentrations of hydrogen and helium in a much bigger cloud that will become, or perhaps already is, a galaxy. As the proto-stars collapse under their own weight, the pressure and temperature inside them build to unimaginable levels. Under these extreme conditions hydrogen atoms are squeezed ever closer and their nuclei fuse together, creating new, heavier elements and, in the process, vast amounts of energy.

The generation of energy within supports the outer layers of the cloud, preventing further collapse for as long as it continues. As nuclear fusion gets underway in the nuclear reactor at its centre, it escapes towards the surface of the seething cloud and eventually appears as light and heat. Thus a star begins to shine, and it briefly illuminates the surrounding cloud of proto-stellar material creating often very beautiful nebulae. These were the subjects of the previous chapter.

But hydrogen fusion is the first of a series of nuclear processes that leads to the creation of a range of ever-heavier elements such as carbon, nitrogen, oxygen and silicon. The more massive the star, the more rapidly this sequence of element creation is traversed. Massive stars are also the most luminous and their radiation carries off a substantial fraction of their substance, sometimes in a spectacular fashion. Such stars are described by astronomers as *mass-loss* stars. Eventually, after millions (for massive stars) or billions of years of stability (for the light-weights) the star runs out of fuel and can no longer support itself. The collapse that first brought it to life is then resumed, sometimes with catastrophic results.

If the star is much more massive than the Sun after its bout of mass loss, it will explode as a supernova, destroying itself and scattering most of its substance far and wide. Stars that lose so much mass in the course of their evolution that they are about the mass of the Sun, or a little more, eject their outer layers as beautiful planetary nebulae. Either way, the blasted remnants of stars, now enriched with the elements of life, are returned to the mixing-pot of the galaxy. These stellar outbursts are themselves often spectacular and are illustrated and discussed in this chapter.

## THE SUN

Because the Sun is the nearest star, we might naturally expect to know more about it than any other. In most ways we do, but it is not normally regarded as a 'mass-loss' star. Such a definition is usually applied to some of the most peculiar, violent and energetic stars, whereas the Sun appears quiescent, stable, and not at all unusual. But all stars lose mass in one way or another. They must do so to survive. Although this chapter emphasises the more interesting and dramatic mass-loss stars, we should start by looking briefly at the Sun, which is neither especially interesting from the mass-loss point of view, nor in any way dramatic, facts for which we have reason to be thankful.

In the case of the nearest star, the nuclear processes in its core change mass (hydrogen) into energy at the prodigious rate of 4 million tonnes *per second*. Though this undoubtedly changes the mass of the Sun, this is not the kind of mass loss we mean in general. However, this energy literally supports the star against gravity, preventing its collapse and eventually leaving the star as light and heat. In the case of the Sun, nuclear fusion is the source of the blue daytime sky, and sustains life on this planet. Also leaving the Sun's seething surface are more tangible particles in the form of charged ions and electrons. These particles are the other, more conventional, form of mass loss. In a cooler, denser environment ions and electrons might come together to form elements and simple compounds, but there is no chance of that in the atmosphere of the Sun.

The particles are shot from the Sun's surface with a temperature of about 2 million K, but there are so few of them that this temperature is really a measure of their velocity, sometimes gusting up to 1000 kilometres per second. And 'gusting' is a fair description,

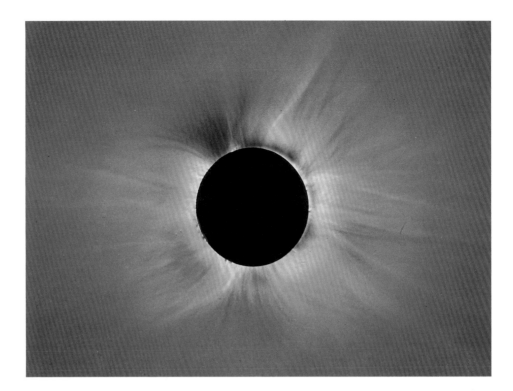

*Fig. 7.1.* Evidence of mass loss from the Sun is best seen during a solar eclipse. This beautiful picture of outflows from the Sun was made in 1981 by Serge Koutchmy.

because this continuous but irregular outflow is known as the solar wind. It is blasted off the surface of the Sun in all directions, though not uniformly, and is the source of the beautiful corona seen during a total solar eclipse (Fig. 7.1). Light-pressure from the Sun increases the velocity of the outward flow and an ever more tenuous solar wind expands to permeate the solar system. Thus all the planets, even the most distant, are in orbit inside the atmosphere of the Sun. The effects of violent events on its surface can be detected at the very edge of the solar system, 6000 million kilometres away from the source.

Fortunately, although the particles are energetic, they are relatively few and far between, and the Earth's atmosphere protects its surface from their continual bombardment. As the rapidly moving particles collide with the upper layers of air they release some of their energy as light and contribute to the night sky air-glow. If the Sun is particularly active and ejects many more particles than usual we see them enter the atmosphere as the aurorae, propelled and directed by the Earth's magnetic field. Until recently, the best detector of the solar wind was the Earth's atmosphere itself and the only hint that the Sun had an outflow came from the streaming appearance of the corona during a solar eclipse. Neither of these phenomena are much use for detecting mass loss from other stars.

## MASS LOSS FROM DISTANT STARS

How, then, do we know that other stars undergo this rather subtle mass loss? The answer is that, in most cases, we do not, certainly not in other Sun-like stars at a Sun-like

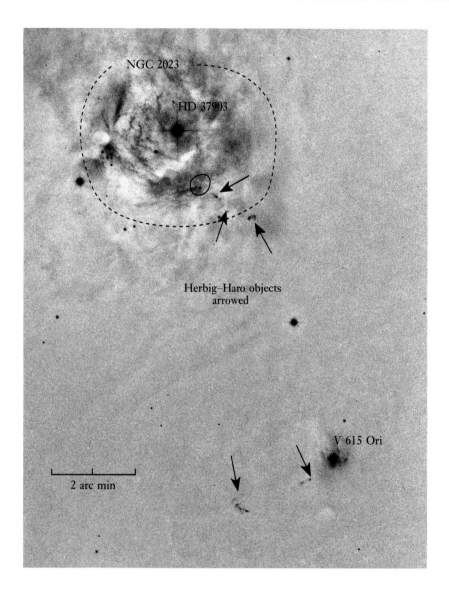

*Fig. 7.2.* An unsharp masked nega-tive print of the red-light plate used for Figs. 5.5 and 5.6 shows the group of tiny Herbig–Haro nebulae (arrowed) associated with the reflec-tion nebula NGC 2023. Thin fila-ments of emission from molecular material surround the star HD 37903. Both were discovered on this AAT plate that was used to make the colour pictures.

stage in their existence. But in some stars, mass loss is not at all subtle, and it is easy to detect at great distances.

Before we consider the 'classical' kind of mass-loss star, which is usually very massive, we should point out that the act of star formation itself can be a mass-loss process, despite the fact that it is normally thought of as the collapse of material onto the surface of a growing star. There are times when the young, still-acquisitive star ejects matter at high speed, often in the form of narrow jets. However, since these events mostly occur in dusty clouds, these jets are hard to find. They were first commented on by Giullemero Haro and George Herbig in 1951–2 and are thus known as Herbig–Haro objects.

Even though the stars that produce them are well hidden, the rapidly moving plumes of matter sometimes travel far enough to escape from the dark clouds from which they spring. As they travel through the dense dust surrounding them, their velocity is

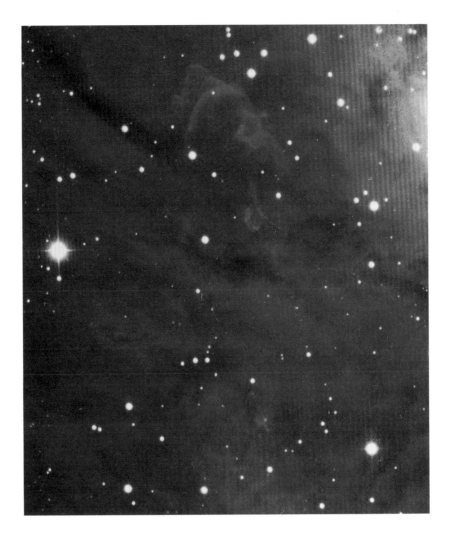

*Fig. 7.3.* Some Herbig–Haro objects form large lacy streaks, which are here seen against a bright background. This group is in NGC 2264, near the Cone nebula.

sometimes sufficient to cause it to fluoresce. The collection of tiny worm-like nebulae at the base of the Horsehead nebula (Fig. 7.2) are typical examples of Herbig–Haro objects. They have distinctive spectra, which make them red in colour, and can be recognised even when the background is not dark. The wispy filaments in Fig. 7.3 are other examples, seen against the nebulosity near the Cone nebula.

## ETA CARINAE

However, weight loss among the stars can sometimes be a much more vigorous business. One of the most intriguing stars in the sky is the southern hemisphere object Eta Carinae. It had long been known to be a star that changed its apparent brightness erratically, and, by good fortune, it was approaching its brightest in the 1830s when Sir John Herschel was observing the largely unknown southern skies from his newly erected observatory near Cape Town, South Africa. He took particular note of this strange star and the nebula alongside it and produced the sketch that appears in Fig. 5.19(a).

At its most luminous, Eta was seen as one of the brightest stars in the sky, outshone

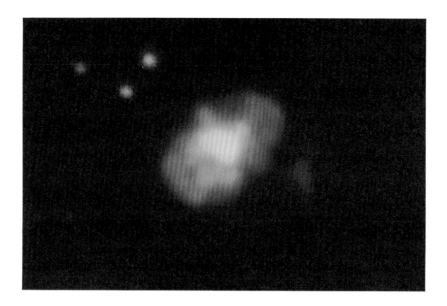

*Fig. 7.4.* The manikin-like Homunculus nebula surrounds the extreme mass-loss star Eta Carinae, effectively hiding the star within.

only by Sirius and the equal of Canopus. It must have been a very bright star indeed, because it is now known that Eta is about 7000 light years away, whereas Sirius is about 9 light years distant and Canopus is less that 200. Today, Eta Carinae seems to have faded by a factor of over 1000, and is well below naked-eye visibility. With the fading of the star has gone the nebula that Herschel was at pains to record for posterity.

Eta's odd behaviour is explained by an extreme bout of mass loss. During Herschel's time at the Cape (in reality, of course, 7000 years before) the star suddenly ejected a substantial fraction of its outer surface, effectively increasing its size and thus its apparent brightness. The newly-brightened star illuminated one side of a nearby dust cloud, which soon became known as the Keyhole nebula, more clearly seen as the central shape in Fig. 5.20. During the next few years, the hot, highly luminous material that the star had ejected gradually cooled and became opaque, partly hiding the light of the star within, which was thus seen to fade.

In the 155 years since Eta Carinae attracted Herschel's attention, the ejected material has continued to expand at several hundred kilometres per second and is now big enough to be visible as a small, bright irregular shape around the star, Fig. 7.4. This was first described by the Argentinian astronomer Ernesto Gaviola in 1950. He likened it to 'a "homunculus", with its head pointing northwest [i.e. to top right] legs opposite and arms folded over a fat body'.* The Homunculus nebula is all one sees when examining this fascinating object visually and it is one of the few objects I have had the chance to have a good look at with the AAT. Even on a night of superb seeing, the slight quivering imposed by the trembling night air and reddish-tan colour accentuate the manikin appearance, making this distant nebula seem to be alive in the eyepiece.

Beyond the visible nebula is another, which is larger and fainter, but still clearly centred on the star. This outer photographic nebula (Fig. 7.5) may be the remnants of an

*Astrophysical Journal* (1950), **111**, 408.

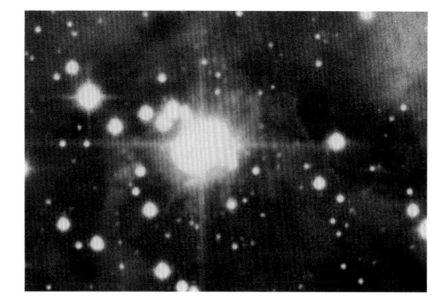

*Fig. 7.5.* A longer exposure of the nebula around Eta Carinae shows its ragged outline and faint outer extensions. It is also a distinctly different colour from the surrounding nebulosity, and the cross-like diffraction spikes reveal the presence of the star itself.

earlier, unrecorded outburst, its tattered shape, curious extensions and spikes suggesting a more vigorous explosion, perhaps a few hundred years ago. The older ejecta from Eta are differentiated by colour and structure from the much smoother background nebulosity and are associated with a complex of wisps and curlicues that hint strongly at the violence of the event.

These layers of opaque material are the condensing outer layers of a star, probably only a fraction of the mass of the underlying body. Even so, assumptions about the composition and opacity of the nebula suggest that it contains enough matter to make 10 stars with the mass of the Sun, and these calculations include only solid material, so the hidden star must be massive indeed. This estimate of mass loss gives a clue to the nature of Eta. It seems to be a particularly ponderous star, perched on the brink of self-destruction.

However, despite this substantial loss of material, a functioning star remains. This depleted star is not merely functional, it is the most extreme star we know of in our Galaxy, an amazing 3 million times more luminous than the Sun. We do not see it as such now because most of the starlight is absorbed by the opaque nebulae around it, especially the Homunculus nebula, from where the star's energy is re-radiated as heat. This tiny nebula is therefore the brightest known at infrared wavelengths, by far the most conspicuous object in the infrared sky.

Though the nebula is almost opaque, some of its light penetrates the hot dust. This produces the cross of the telescope's diffraction spikes in Fig. 7.5, an artefact seen on all deep images of bright point sources (i.e. stars) obtained with reflecting telescopes such as the AAT. Though the sparkle of Eta is dimmed by the products of its own instability, the Carina nebula contains other jewels. Within the region covered by Herschel's drawing, reproduced in Fig. 5.19, and in the closer view seen in Fig. 7.5, are some of the most massive and most luminous stars in the Galaxy, some only a little less brilliant than Eta itself. They seem faint to us because they are 7000 light years away, but these stars *are* extreme. They are also true 'mass-loss' stars in the sense understood by most astronomers. Such stars do not usually draw attention to themselves by extreme variability or by the ejection of rapidly expanding, highly visible nebulae, but their surface temperatures and distinctive spectra reveal their unusual nature. This was first noted by the French astronomers Charles Wolf and Georges Rayet in 1867, and they are now known as Wolf–Rayet stars.

They are unusual because they are very massive, up to 40 times the mass of the Sun. They are also very hot, with surface temperatures of many tens of thousands of kelvins, and their spectra show that they are shedding mass at an astonishing rate. One solar mass of material is ejected every few thousand years in the most extreme examples. Such behaviour cannot long be sustained and thus such stars are rare: fewer than 200 are known in our Galaxy, despite their visibility.

These stars are condemned by their mass to lose sufficient of their substance to become less luminous and just a little cooler. Despite the vigour with which such stars run their lives, visible manifestations of steady mass loss are rare. But such massive outflows of energy and mass cannot always go completely unseen, and occasionally Wolf–Rayet stars produce nebulae, either because their mass loss is sporadic, as with Eta Carinae, or because the outflow collides with surrounding, relatively stationary material, the shock of the interaction producing a visible nebula.

## ENORMOUS STARS AND MASSIVE MASS LOSS

Some of the best examples of nebulae associated with stellar outflows are in the LMC, a companion galaxy of the Milky Way, 170 000 light years away. To those who live in the southern hemisphere, the LMC is perfectly placed to probe the anatomy of a modest-sized irregular galaxy in considerable detail. This is of interest because the LMC is in many ways quite unlike the other galaxy we are most familiar with, the Milky Way. One of its curious features, shared with many relatively low-mass galaxies, is an interesting assortment of huge, roughly spherical bubble-like nebulae that surround groups of very hot stars.

These objects have few obvious counterparts in the Milky Way, except perhaps the incomplete shell of Barnard's Loop, which partly surrounds the Orion star-forming region, but they are sometimes found in galaxies like the LMC. One of the biggest LMC mass-loss bubbles is seen in Fig. 7.6. Henize 70 is almost 400 light years across, 100 times the distance from the Sun to the nearest star. At its centre is a group of hot stars similar to those we saw in Carina; they are responsible for the production of the energy that has created the bubble, which is the visible surface of a thin-shelled sphere. The light of the nebula has almost certainly not originated in the stars themselves, but from the interaction of their 3000 kilometres per second stellar winds with the almost stationary interstellar medium within the LMC.

This group of stars dominates an enormous volume of space. If the density of stars in that part of the LMC were the same as it is near the Sun, millions of them would be inside the Henize 70 sphere. If any of these stars had planets, their atmospheres would almost certainly have vanished, destroyed by the unremitting interstellar blizzard of energetic particles from their distant but unruly cousins.

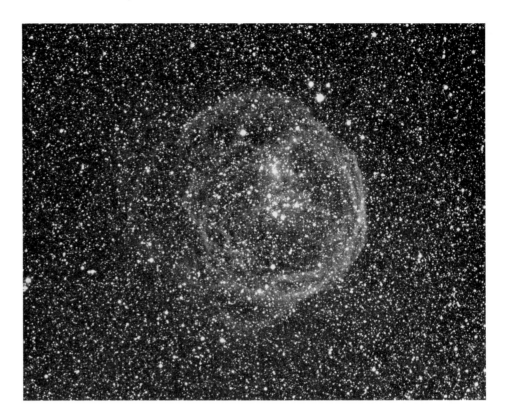

*Fig.* 7.6. Henize 70 is an enormous stellar wind-blown bubble surrounding a group of very luminous stars in the LMC.

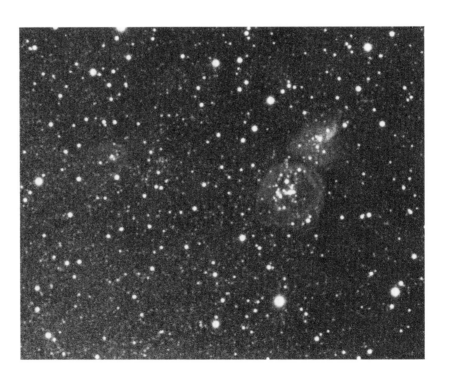

*Fig.* 7.7. Bubble nebulae are common in low-mass, gas-rich galaxies. These examples are in NGC 6822, shown in its entirety in Fig. 8.12.

Similar bubble nebulae can also be found in other nearby LMC-like galaxies such as the nearby irregular system NGC 6822 (Fig. 7.7). The only clear, large example in our own Galaxy is named Barnard's Loop, after its discoverer. It is a very faint arc partly

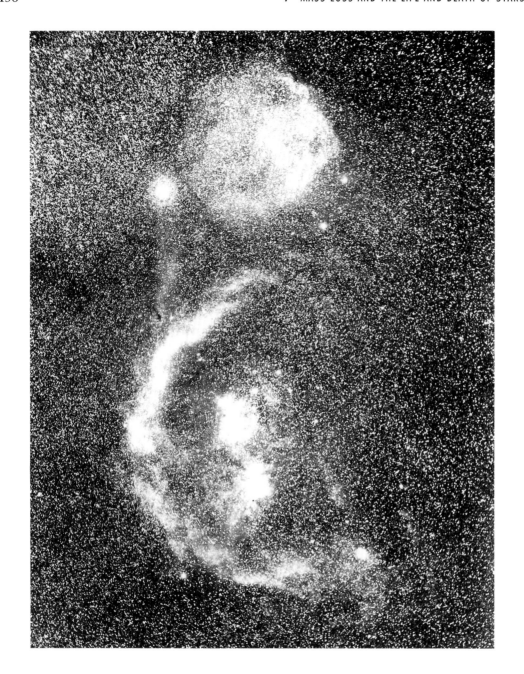

*Fig. 7.8.* Barnard's Loop is centred on the star-forming activity in Orion. The two bright blobs near its centre are the Orion and Horsehead nebulae.

encircling the Orion and Horsehead nebulae and their associated stars. I made the picture reproduced as Fig. 7.8 with an astronomical plate exposed in an ordinary plate camera. Because the loop is so close it seems enormous, nearly 20° from end to end, and too big to be photographed with any normal telescope.

But the LMC is also relatively close, and we can therefore see its detailed structure. Because the LMC is a gas-rich, irregular galaxy, an environment where wind-blown bubbles seem common, we can find many of them at a variety of stages in their development. Fig. 7.9 shows Henize 44, a cavity blown in the interstellar medium by a group of hot stars. The shell-like structure is faint but the irregular

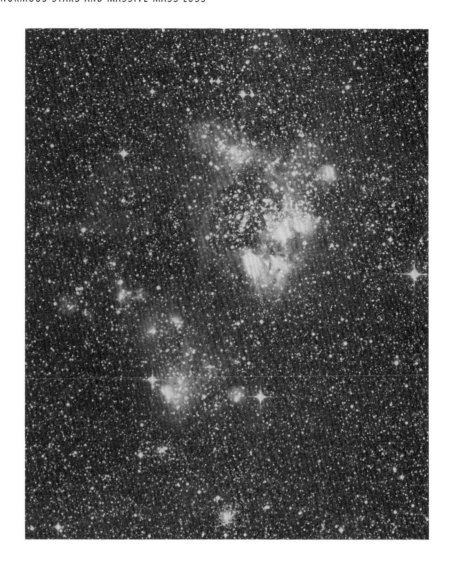

*Fig.* 7.9.   Henize 44 is one of many regions in the LMC where hot stars have cleared away the surrounding nebulosity.

outline is almost certainly due to variations in density or motion in the surrounding nebulosity.

Although rare, spherical nebulae associated with hot stars are found in the Milky Way. One such is the relatively small bubble associated with a fairly insignificant-looking Wolf–Rayet star, tangled in the nebulosity seen in Fig. 7.10. This one has created a roughly spherical blueish shell of material and is associated with a wispy red nebula, NGC 2359. It is possible that the nebula also originated from the star and is the cooling remains of an earlier phase of ejection, still illuminated by the energetic star at the centre of all the activity. A similar scenario fits another bubble nebula, *the* Bubble nebula, NGC 7635, which surrounds a very hot star that has blown a delicate 'blister' in a cloud of glowing hydrogen (Fig. 7.11).

Other stars have episodes of mass loss so vigorous that they eject large amounts of material, like Eta Carinae. Sometimes these are surprisingly symmetrical, as in Fig. 7.12. The lobes of the nebula around the star HD 148937 are so conspicuous and apparently

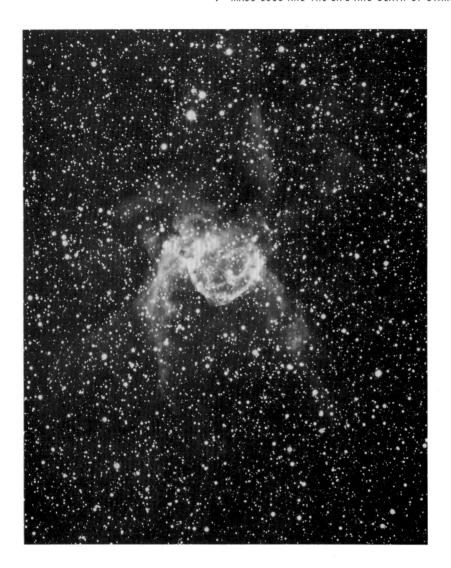

*Fig. 7.10.* Stellar winds and mass loss from a Wolf–Rayet star near its centre has produced the blueish bubble in the NGC 2359 nebula.

distant from the central star that they were given two separate catalogue numbers by the compilers of the *New General Catalogue,* who did not have photography to guide them. The symmetrical bipolar structure of NGC 6164–65 is common in outflows from stars, and in this case the shells of nebulosity seem to have been ejected from the poles of the star, which is rotating unusually rapidly and is over 4000 light years away.

The nebula itself extends six light years from end to end and contains enough material to make two stars with the mass of the Sun. As with Eta Carinae, this star has undergone more than one episode of sudden mass loss and is an unusually active star, perhaps in transition from one kind of Wolf–Rayet star to another. A deeper, wide-angle picture (Fig. 7.13), which I extracted from the same set of AAT plates, has NGC 6164–65 at its centre, but shows signs of an earlier outburst in the faint, crenellated nebula around part of the photograph. This outer nebula is over 30 light years from end to end, and on very deep pictures an even bigger nebula can be seen beyond it, which indicates that HD 148937 influences a volume of space 150 light years across.

*Fig. 7.11.* The Bubble nebula, NGC 7635, is about 8000 light years away and is a cavity several light years in diameter blown by the extremely hot star near its northern rim.

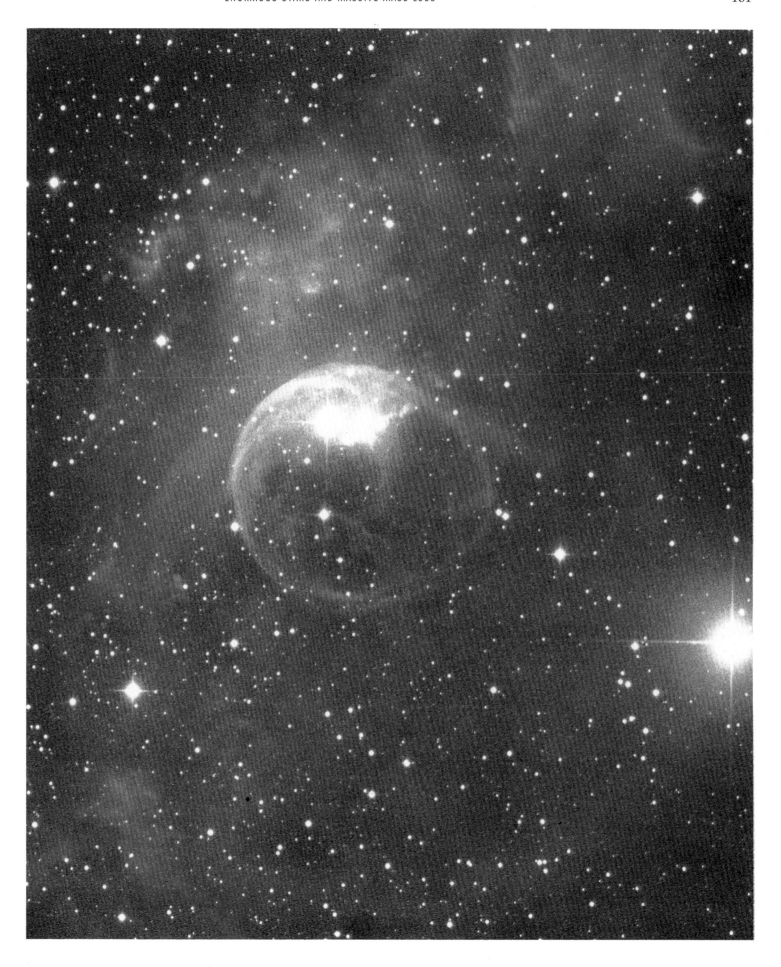

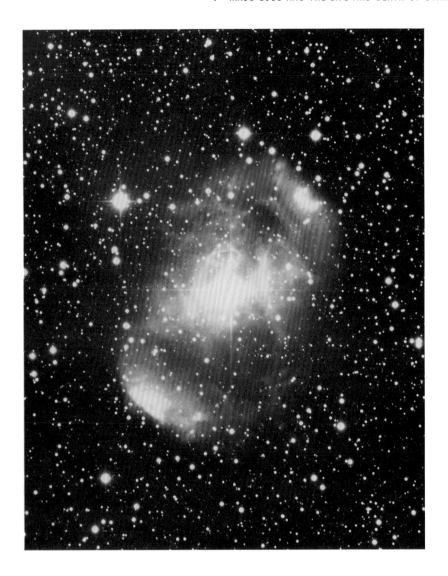

*Fig. 7.12.* The remarkably symmetrical nebulae NGC 6164–65 originate from a very bright multiple star HD 148937 in the southern constellation of Norma.

That there are so few Wolf–Rayet stars and even fewer with visible nebulae indicates both that their precursor stars are rare, and also that the energetic, extrovert behaviour that draws attention to them cannot go on for very long. But not all mass-loss stars are Wolf–Rayet stars, some are merely massive. As they squander their substance both by energy production within and by driving matter away from their boiling surfaces, these stars change weight quickly. In less than a million years a star can have changed from an unruly youth of perhaps 30–40 times the mass of the Sun and subject to violent outbursts to a more mature and stable 10–15 solar mass star, still brilliant, but much more predictable.

The life history of such stars is now fairly well understood. As the central store of hydrogen is converted into heavier elements, energy production in the core briefly ceases and the core shrinks. But hydrogen continues to fuse into helium in a sort of shell around the core of the star, and energy from this reaction causes the outer layers of the star to expand. This is not a modest increase in size. If one such massive star, with a

*Fig. 7.13.* A deeper image of the region around HD 148937 shows evidence of an earlier active phase in a faint, crenellated nebula at a considerable distance from the central star.

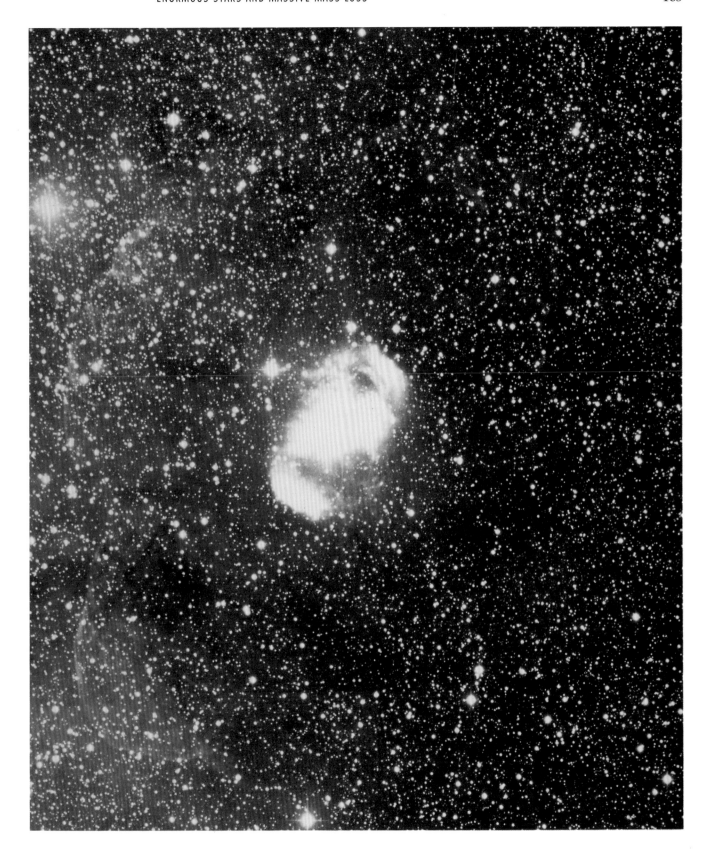

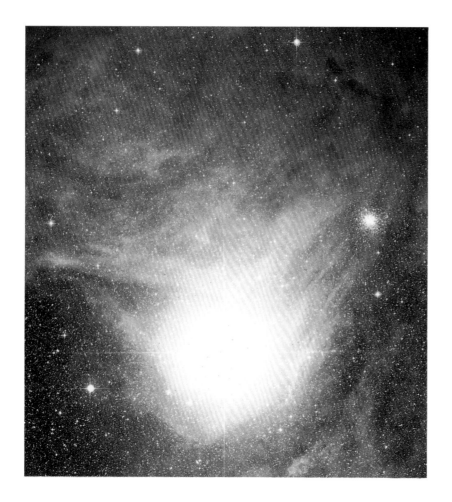

*Fig. 7.14.* The bright supergiant star Antares illuminates a large, faint, irregular nebula that it has created by loss of particles from its cool, distended surface.

diameter a few times that of the Sun, were to begin its expansion at the centre of the solar system, both the Earth and Mars would soon be engulfed in its outer envelope. As the star's surface expands it cools from tens of thousands of kelvins to a more moderate 3000 K or so. The star has metamorphosed into a 'red' giant (or supergiant, depending on its mass).

## RED GIANTS AND PLANETARY NEBULAE

Elsewhere in this book I have questioned the term 'red giant'. Such stars are certainly giant, but they are not red, no redder than a domestic filament lamp. However, despite the dramatic lowering of temperature that arriving at the red giant phase implies, such stars continue to lose mass. In the cool, extended envelope of a gigantic star, those atomic fragments that were too agitated to combine in the Sun's million kelvin atmosphere condense to form recognisable elements and compounds. Most of them are simple, stable gassy molecules such as oxygen, nitrogen, carbon dioxide, carbon monoxide and water. Others are refractory solids such as silica, carbon and oxides of calcium and titanium. The solids form as tiny particles in the outer atmosphere of the swollen star, much

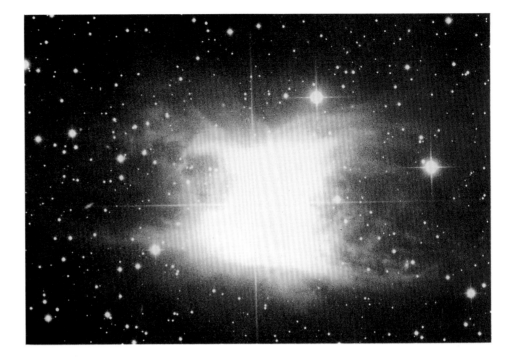

*Fig. 7.15.* IC 2220 is the Toby Jug reflection nebula around a cool, bright star. This is a deeper picture than that reproduced in Fig. 5.15 and shows that the faint, outer lobes of the nebula have a remarkably symmetrical bipolar structure.

as smoke appears above the flames of a fire. And like smoke, these particles affect the colour of the light from the star beneath, absorbing and scattering blue light, making a yellowish star seem a deeper shade of orange.

Because the star is so huge, gravity at its surface is low and the smoky particles drift away as if wafted by a gentle breeze. Sometimes the motes of dust are so numerous that they reflect the light of the star from which they came and can be detected as reflection nebulae. The wispy yellow cloud around the conspicuous naked-eye star Antares in Fig. 7.14 is one of the best examples of such mass loss. All 'red' giants lose mass in this way, but few are such prolific polluters of the space between the stars as Antares.

Another example is seen in Fig. 7.15, which is a reflection nebula seen elsewhere in this book as IC 2220 (Fig. 5.15). It surrounds another red giant star, which is not bright enough to have a popular name. The star appears in catalogues as HD 65750 and is believed to be about 1000 light years away, twice the distance of Antares. Its butterfly-like cloud of dusty particles has reminded some astronomers of the shape of an English Toby Jug, a kind of fancy drinking mug with a large, curved handle. I made this picture over 10 years ago for a previous publication, but improvements in technique encouraged me to work on the plates again and my latest picture is much deeper. It shows a bipolar structure like a bow-tie, a form characteristic of many stellar outflows.

In places the clouds of dust associated with a star can be so dense and patchy that they occasionally hide the source of the light inside it, producing what seems to us to be an irregularly variable star, though the nebula itself does not vary in brightness. Similar random variations in the light of other cool giants are thought to be due to the passage of smaller but otherwise unseen dust clouds drifting across the face of the star. The bright

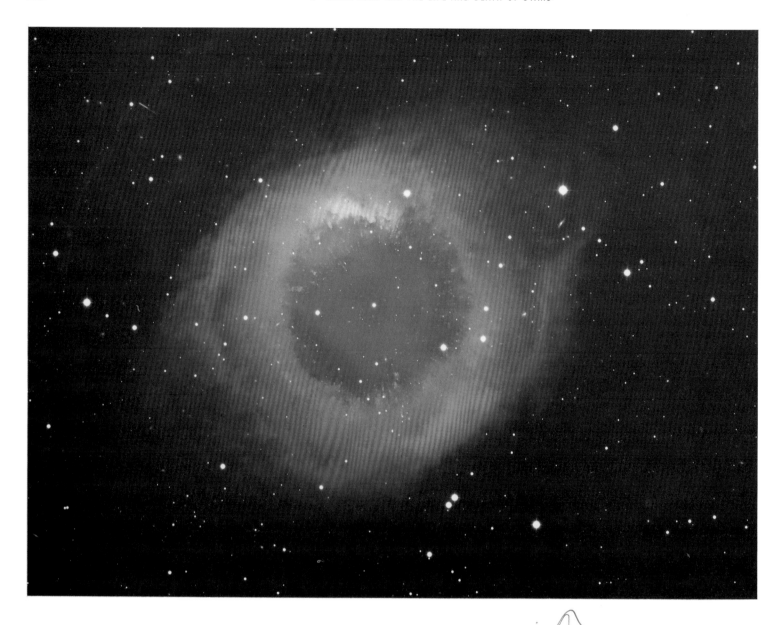

yellow star in NGC 3293 (Fig. 5.12) is such a variable, but very few red giants have large nebulae like HD 65750 and Antares.

Towards the end of their lives, massive stars may undergo several stages of mass loss, reflecting a series of internal rearrangements as their energy source is steadily depleted. Their appearance may change from cool, distended and yellowish to hot, blue and compact over a few tens of thousands of years, and back again, perhaps several times in what amounts to a mere blink of an eye on the cosmic time-scale. Thus the stars become un-stable and sometimes pause briefly in a kind of intermediate state, pulsating on a time-scale of days or weeks as the surface of the star shrinks and expands, warms and cools in response to cyclical changes in its interior.

Eventually, this instability comes to a close, but the manner of the final cataclysm

*Fig. 7.16.* The Helix nebula (NGC 7293) is the nearest of the planetary nebulae and the largest, at over half a degree across, about the same apparent size as the full Moon. The nebula is excited by a white dwarf star at its centre, the exposed core of the star that produced the nebula.

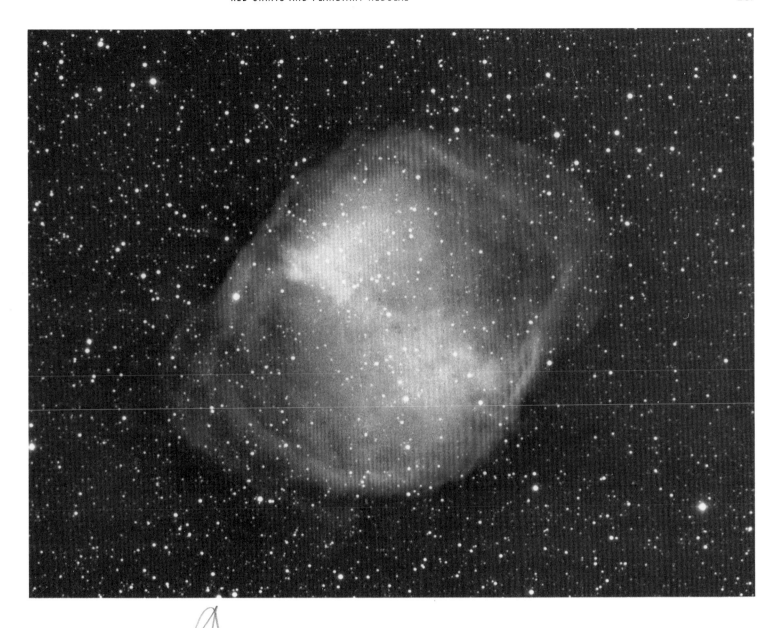

Fig. 7.17. The Dumbbell nebula, M27, derives its name from the visual appearance of its brighter parts in a small telescope. This picture, made from plates taken with the Isaac Newton Telescope, shows some clumpy internal structures and extensions too faint to be seen by eye, but the symmetry implied by the popular name is quite obvious.

was determined millions of years ago when the star formed. Massive stars end their brief, brilliant lives as supernovae, and we will look at these later in this chapter. Stars of more modest mass signal their end with a gentler outburst.

Low or moderate mass stars such as the Sun, or those high mass stars that have undergone substantial mass loss will also swell to become red giants, shedding vast amounts of solid particles around them. At the end of this stage, quite suddenly, they throw off their outer layers and transform themselves into planetary nebulae. This name was given to these distinctive objects by Sir William Herschel, who discovered them 200 years ago: he thought their circular shapes to be reminiscent of the disks of planets as seen through a small telescope. They often appear as beautiful, bright circular nebulae around faint central stars. Herschel initially believed that he was seeing solar systems in the mak-

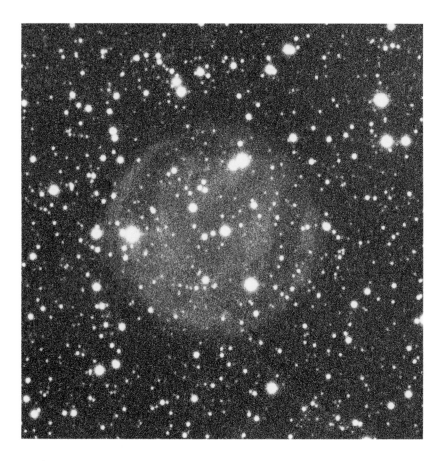

*Fig. 7.18.* This is the first colour photograph of a very faint planetary nebula, Ack 277-03.1. It shows the same kind of symmetry as the Helix and Dumbbell nebulae, but without any sign of the distinctive radial streaks.

ing, distant Suns surrounded by luminous material condensing into planets, and named them 'planetary nebulae'. Instead, what he was seeing was the demise of a star and the incineration of any planets it might have had.

These beautiful nebulae are the outer layers of a star, blown off by a burst of nuclear burning in a 'shell' of hydrogen surrounding its core. In many cases, all the material in the shell ignites at about the same time and the star is blown apart in an orderly fashion. These nebulae thus often have a satisfying but complex symmetry as in the Helix nebula in Fig. 7.16. The removal of the surface layers of the star reveals the much hotter core beneath, and the central stars of planetaries are some of the hottest known with temperatures of 50 000 K or more. At such elevated temperatures most of the star's energy is radiated in the ultraviolet part of the spectrum, and for a short time this excites the surrounding nebula making it fluoresce like an advertising sign. But what is left of the star, now technically but prosaically described as a white dwarf, is deprived of the overburden of fuel that kept it alive for so long.

The ejected nebula expands into the region around the star, perhaps encountering the solid particles sloughed off during the earlier red giant phase. For a short time the colours of the planetary nebula tell of the composition of the recently shed outer layers of the original star. The green at the centre of the Helix nebula is mainly the result of excited oxygen atoms, while some of the red is from hydrogen. Spectra of these nebulae also show

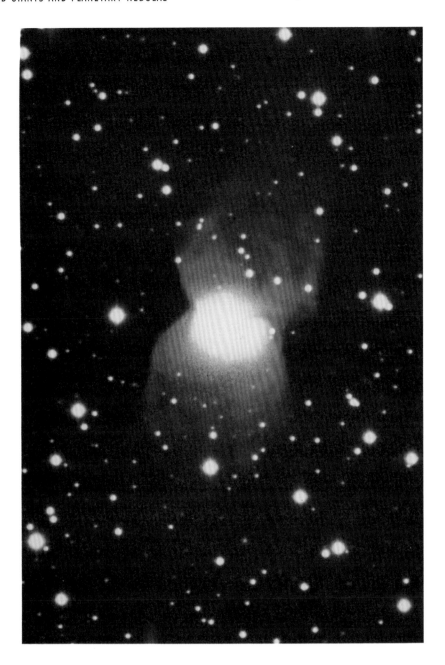

*Fig. 7.19.* The sharply-defined V-shaped boundaries of planetary nebulae such as NGC 2346 are very common and probably reflect the circumstances of their formation and the orientation at which they are seen.

sulphur, nitrogen, carbon, chlorine and iron, as well as the inert gases helium and neon. Of these only hydrogen and helium were present in the early Universe. It is from such stars that the elements of life are returned to the Galaxy, perhaps to be incorporated into later generations of stars and planets for which the plans are not yet drawn.

But for the naked core of the central star regeneration is over. It has now entered the white dwarf phase, and since it no longer retains the lighter elements necessary to provide the energy to keep it shining, it cools. Eventually this stellar remnant will disappear from sight at visible wavelengths. Long before, the radiation of the ultraviolet light that makes the gas visible will have ceased and the nebula around it will eventually disperse.

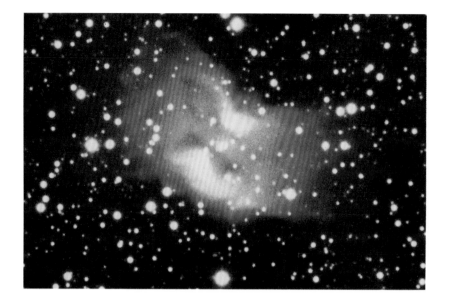

*Fig. 7.20.* Not all planetaries have perfect symmetry, as seen here in NGC 2899.

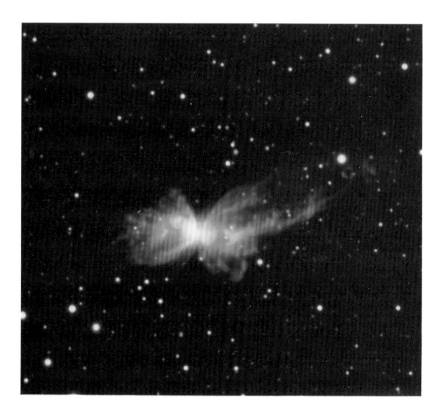

*Fig. 7.21.* NGC 6302, the Butterfly (or Bug) nebula, has an unusually high velocity of expansion and does not seem to have a central star.

Large numbers of planetary nebulae are known, but only a few are as spectacular as the Helix and many are much fainter. One that is not is the well-known Dumbbell nebula, M27 (NGC 6853), pictured in Fig. 7.17. William Herschel initially believed this to be 'a double stratum of stars of a very great extent' and sketched what he thought to be its three-dimensional structure in his notebook in 1784. Though we now know that planetaries are glowing gas, their form is still difficult to understand simply by looking at pictures, though colour images help to disentangle the various components.

A much fainter but similarly beautiful example is seen in Fig. 7.18. This is listed as Ack 277-03.1 (after Agnes Acker, the cataloguer) and is so faint that I had to use rather heavy-handed photographic amplification to extract the colour picture, which is why the image is so grainy. Many other planetary nebulae, such as NGC 2346 (Fig. 7.19), are brighter, and show remarkable symmetry, while others, such as NGC 2899 (Fig. 7.20), have similar shapes but without the symmetry. Continuing the theme of symmetrical outbursts, the unusual planetary nebula NGC 6302 (Fig. 7.21) has been likened to a butterfly or, less poetically, a bug. It is probable that the magnetic fields of the original star have dictated the form of this star's last gasp. Curiously, the central star is not visible here, but the spectrum of the nebula indicates that it is excited by a particularly hot white dwarf.

## SUPERNOVA!

Returning to the more massive stars, they too end their lives in mass-loss events, exploding as supernovae, the most dramatic cosmic climax. It is not only the drama of their end that makes these stars interesting. The study of supernovae brings together two branches of physics that at first sight seem to have no connection. The supernova event illuminates the fundamental link between the particle physicists, who study the building blocks of matter on the minutest scales, and the astronomers, who seek to understand the functioning of the stars themselves. Both find fresh sustenance and much common ground by examining the remnants of those stars that have exploded.

To an astronomer, especially a theorist, who has wrestled with the unobservable, almost unimaginable events that occur in the deep interior of stars, a supernova is an unparalleled chance to glimpse something of the truth. The particle physicist similarly sees reactions occurring under conditions so extreme that they can never be duplicated in a terrestrial laboratory. The supernova explosion and its immediate aftermath provides an avalanche of facts with which both can refine – or reject – their theories.

But in witnessing the destruction of a star we are also watching regeneration. Supernova explosions are the only source we know of the heavier elements that we find around us. The silver, gold and platinum that we treasure, and the serviceable copper and zinc, tin and weighty lead and uranium – all were formed inside massive stars that lived briefly and died violently before the Sun and its retinue of planets appeared.

During the explosion, which ironically starts with an internal collapse, most of the star is suddenly blasted away from the core. The expanding, enriched material is unbelievably hot and travelling outwards at thousands of kilometres a second. It will eventually cool and condense, adding its share of heavy elements to the cold dark clouds in the host galaxy, together with the lighter froth blown out of planetary nebulae.

In its early phases, the supernova also generates cosmic rays, radiation of enormous energy and penetrating power that rattles around the Galaxy with the speed of light. Occasionally an energetic cosmic-ray photon dives deep into a dark, dusty cloud where little else penetrates, its energy encouraging chemical reactions between the frozen ingredients. Such violent prodding contributes to the complexity of the simple molecules that already exist there as the cooled left-overs of mass-loss stars. These substances are themselves important, at least to us, since we too are made of complex molecules.

The shock wave that travels out from the supernova stirs up these quiescent but unstable clouds, eventually triggering the formation of more stars, and possibly even driving the wave of star formation that powers the progress of spiral arms in gas-rich galaxies such as the Milky Way. Supernovae are therefore not just interesting random events that nourish the minds of theorists, they are the dynamic force that drives star formation in a galaxy and, in the widest sense, contributes to their richness and diversity.

Supernovae are fairly uncommon events because the massive stars that spawn them are themselves quite rare. While there may be a few supernova explosions in a typical human lifetime among the 100 thousand million stars that populate the Milky Way, most of our Galaxy is hidden from us by dust, so that they seem to be even rarer than they are. In a Milky Way-like galaxy that we see face-on, such as M83 (Fig. 4.10) we witness them largely undiminished, and M83 has had more supernovae than almost any other nearby galaxy; at least 11 have been seen in the last 50 years.

For a few weeks the exploding star can be as bright as the galaxy it inhabits and can be seen over enormous reaches of space. None had been seen nearby since the invention of the telescope in 1609, but all that changed when supernova 1987A was seen to appear in the LMC. An insignificant twelfth magnitude blue star, which had been previously studied and catalogued as Sanduleak −69°202, brightened over 2000 times as news of a catastrophic internal collapse reached its surface and blew it asunder. Of course, all this happened 170 000 years ago, but news of it, carried by the pulse of light (and some invisible particles, which actually arrived before the light but were not recognised until later) was first seen by an astronomer working in Chile on 27 February 1987.

This event was one of the most important to occur in astronomy for decades, and perhaps the most important ever for the relatively few southern hemisphere observatories that could observe its aftermath. Pressure for telescope time everywhere in the southern hemisphere was intense as schedules prepared months before were hastily reorganised and observing strategies contrived to make the best possible use of what was likely to be a once-in-a-lifetime opportunity. Director's time, 'service' time, even normally sacred engineering time were sacrificed for the supernova. Luckily, I had a night on the telescope in early March, during a blizzard of rescheduling that became an avalanche.

Fig. 7.22 was taken then, and accurately records the colour of the star about two weeks after the event; the expanding shell of ejecta, containing most of the mass of the blue star, had cooled to a deep yellow. The star only appears large on our picture because it was so bright. Its image is no bigger than any of the other stars in the LMC, but it contains so much light that even a tiny proportion of it scattered in the photographic emulsion and by the telescope optics is recorded as a swollen image.

In Fig. 7.22 I have superimposed a negative picture of the same field, derived from a plate taken in 1984, three years before the supernova appeared. It is superimposed in register with the supernova picture and shows the curiously distorted image of Sanduleak

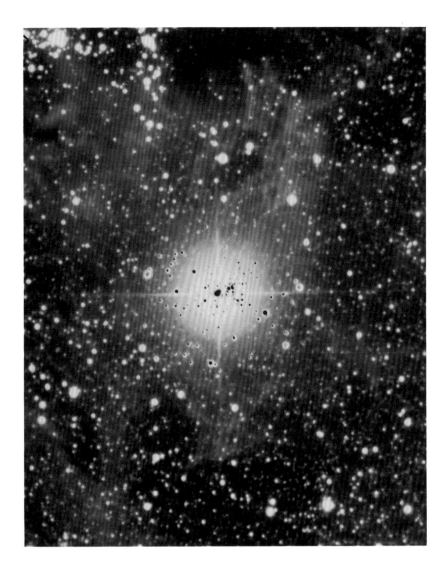

*Fig. 7.22.* Supernova 1987A in the LMC was photographed with the AAT in March 1987 as it neared its brightest. Overlaid in register on the colour picture is a negative image made from a plate taken some years before the supernova appeared. The image of the star that was to become a supernova is confused with two others in the same line of sight and so appears non-circular.

−69°202, the star that exploded. It only seems distorted because the star's light is blended with that from at least two other, fainter stars that happen to be in the same line of sight. In the first days of excitement about the supernova there was some uncertainty about which of these had exploded, but measurements of this plate made by Graeme White confirmed the Sanduleak star as the precursor. This star had been classified in a routine survey of bright LMC stars by Nicholas Sanduleak and his colleagues some years earlier and so became the first supernova to be identified with a star that was known before the event.

Four years after it flashed into prominence, I photographed the field again. Now, the glowing embers of the fast-fading supernova are visible as the slightly fuzzy red 'star' in Fig. 7.23. The two blue stars on either side of it are seen as individuals in this very short exposure. They are responsible for distorting the image of Sanduleak −69°202 in the much longer exposure used to create the negative overlay in Fig. 7.22. These stars are so far untouched by the explosion, but are almost certainly in the LMC and were witness

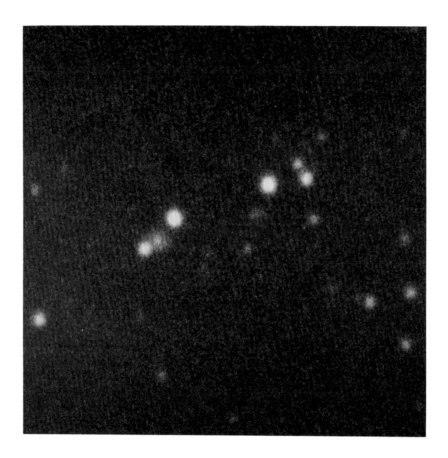

*Fig. 7.23.* A feeble, slightly elongated red glow remains where supernova 1987A exploded four years previously. The bright blue star that was here has vanished and the two stars that complicated the pre-explosion image in Fig. 7.22 are now seen as individuals.

to the destruction of the Sanduleak star 170 000 years ago. The picture shows the stellar remnant to be elongated, roughly east–west, in the same way that the famous Hubble Space Telescope picture of the ellipse around the supernova is elongated.

Supernova 1987A is the most recent addition to the rather short list of supernovae with nebulous remnants. The best known of these is the Crab nebula, the expanding remains of a star that was seen to explode in 1054 and was recorded as a 'guest star' by Chinese astronomers. In that part of the sky now there is a nebula that has probably been studied more than any other. It has been said that much of modern astronomy can be divided into two parts: the study of the Crab supernova and the study of everything else. By chance, this well-known nebula is also number one in Messier's famous catalogue.

My pictures of this fascinating object are some of the few in this book that were not made with telescopes of the Anglo-Australian Observatory. Fig. 7.24 was made from plates taken in the 1960s on the 5 metre Hale telescope on Mt Palomar, when it was known as the 200 inch. The Crab nebula can be seen from Australia, and the AAT can slew to the far north to reach it, but it is never very far above the horizon, and is not accessible for long. At the extreme zenith angles needed to reach such a northerly declination the light from the object passes through an enormous thickness of atmosphere, so the image quality is therefore usually poor. Even when it is not poor, image quality overhead is likely to be very much better, so for one reason or another I have never photographed the Crab nebula. The Palomar plates were kindly made available to me through the good offices of Bob Brucato

*Fig. 7.24.* M1, the Crab nebula is the expanding remnant of a star that was seen to explode almost 1000 years ago. Apart from the nebula, the explosion left behind a rapidly spinning neutron star, the Crab pulsar.

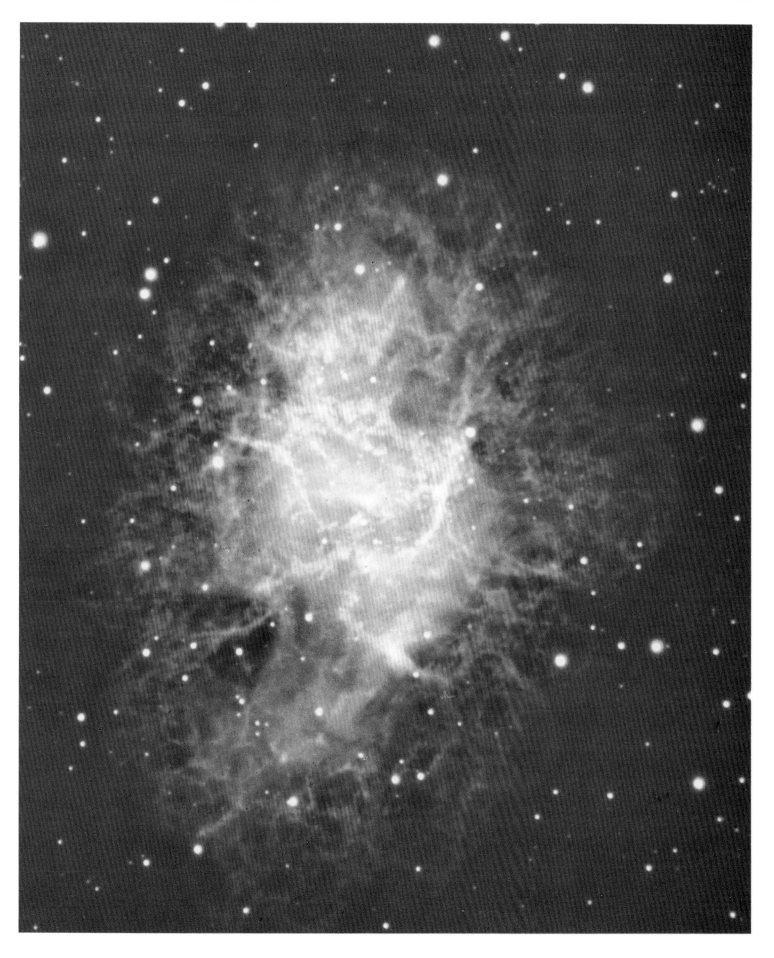

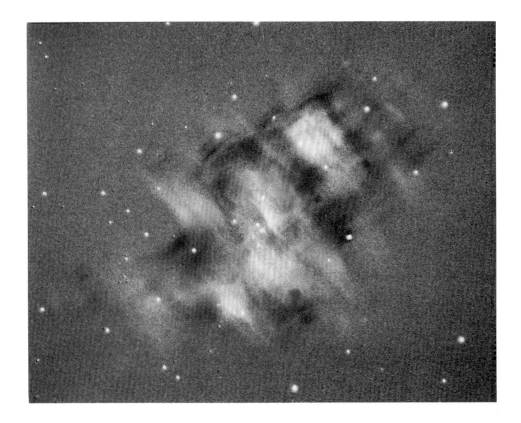

*Fig. 7.25.* The smooth blue-green component of the Crab nebula seen in Fig. 7.24 is the strongly polarised synchrotron nebula. This false-colour picture combines information from four plates taken with polarising filters set at different angles and reveals a surprising symmetry not seen in unpolarised light.

and Neill Reid, with the encouragement and help of Jay Pasachoff, always keen to have new pictures to illustrate his well-known textbooks.

The colour picture reveals the two main optical components, the network of red filaments and a diffuse, blueish nebula. The filaments are an emission nebula predominantly made of hydrogen, the unburned remains of the star that exploded. The hydrogen is laced with the heavier elements that one expects to find in a mature star, especially one that has destroyed itself. The wisps of gas have a mass about that of the Sun, and are induced to glow by the other part of the supernova remnant, the massive, compact pulsar that was the core of the exploding star. This tiny object is only 30 kilometres across, yet it too has a mass of about that of the Sun. Spinning at the astonishing rate of 33 times a second, the stellar remnant still emits enough energy to excite a substantial nebula.

By contrast, the almost blueish cloud is a synchrotron nebula that seems remarkably smooth in texture and uniform in colour, almost quiescent. But the nebulosity is the result of electrons gyrating in a powerful magnetic field, losing energy in the form of light (and radio radiation) as they do so. However, the unusual origin of this light also makes it polarised, and the direction of vibration of the light waves tells us about the orientation of the magnetic fields in the nebula. The relative amounts of polarisation in various directions can be recorded photographically using a polarising filter, in the way one would use such a filter to cut down the reflection from the surface of the sea or from a glass window. To analyse the polarisation structure, it is normal to take separate plates with Polaroid set to four different angles 45° apart.

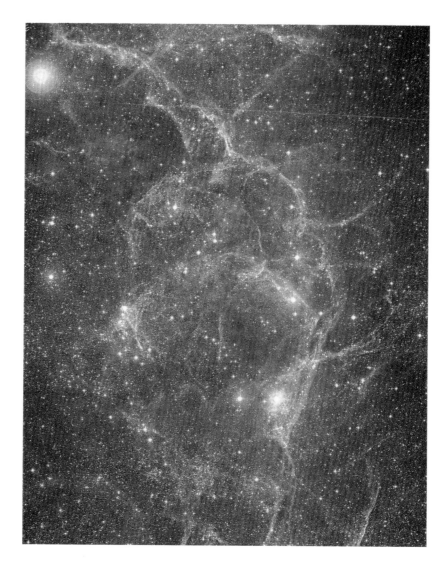

*Fig. 7.26.* The north-western quadrant of the Vela supernova remnant marks the region where the expanding outflow from the exploding star meets the more or less stationary interstellar medium. The variable density of the tenuous gas between the stars produces a very thin, strongly wrinkled nebula.

The resulting pictures are at first sight rather uninteresting, but the plates from Palomar arrived in Australia at about the time that I was experimenting with the image subtraction technique I used for the supernova 1987A light echo images. I decided to do something similar with the polarisation plates. I made a positive copy of one of the negatives and, by overlaying it in register, subtracted it from (or cancelled it with) the three others one at a time, thus producing three 'difference' images that contained the polarisation information. It was then a simple matter to combine these in the normal way to make a three-colour, false-colour picture that contained all the information from the four original plates. The result was the remarkable image seen in Fig. 7.25.

A closer, but more ancient supernova also has a pulsar, but shows no signs of a synchrotron nebula. A star exploded in the southern constellation of Vela about 10 000 years ago. Its expanding shell is now about 10° across, and very faint. Special image enhancement techniques were needed to extract a colour image of the brightest nebulosity from UK Schmidt plates. The Vela supernova remnant is seen in Fig. 7.26,

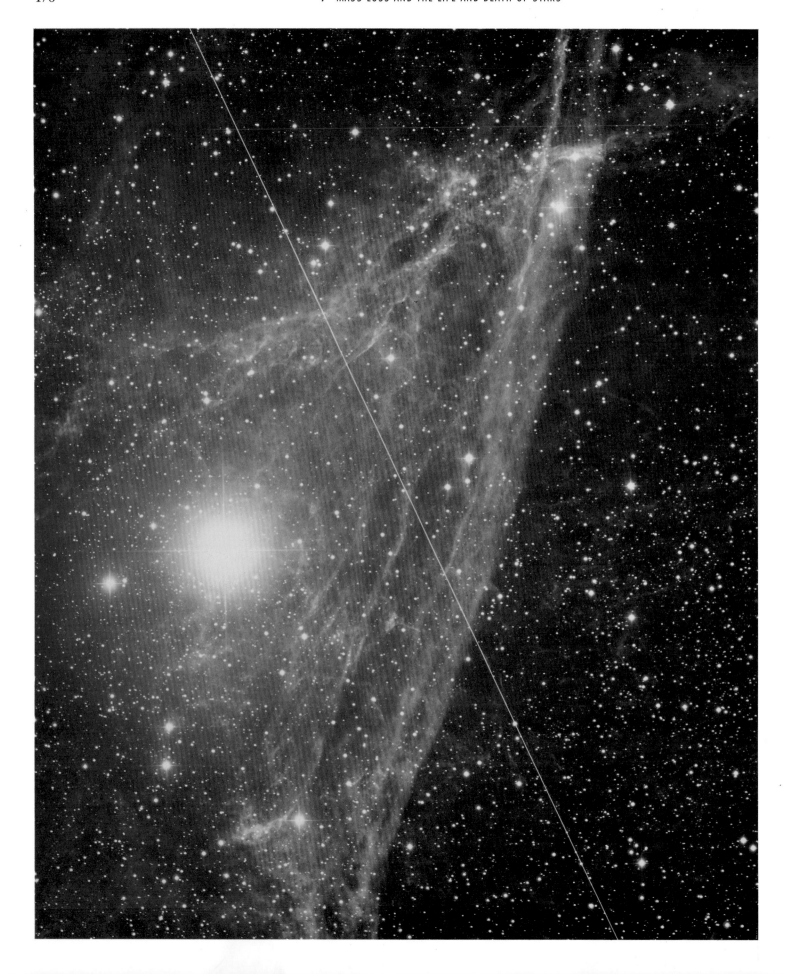

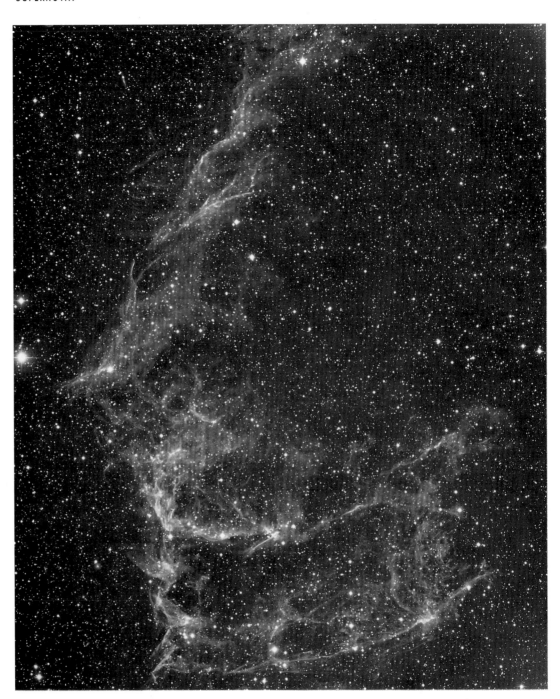

*Fig. 7.28.* NGC 6992 is part of the Veil nebula or Cygnus Loop, a bright supernova remnant about 2500 light years away.

*Fig. 7.27.* More detail is seen in this AAT picture of a small part of the Vela supernova remnant. The straight line across the picture is the track of an artificial satellite that passed through the field during the green-light exposure.

clear evidence of the vigour with which exploding stars sweep up material from between the stars as the wave of energy they produce races through the galaxy, stirring up the relatively stationary interstellar medium. The blue shock front of the nebula is more evident in Fig. 7.27, a new picture I made recently from new AAT plates, and in Fig. 7.28, part of the Cygnus Loop, or Veil nebula, a photograph I made from plates taken with the Isaac Newton Telescope. This supernova remnant is about 2500 light years away and is bright enough for this part of it to appear in the *New General Catalogue* as NGC 6992.

The energy from exploding stars does more than produce delicate nebulae. The enriched materials they return to the galaxy provide the substance from which new stars and their associated planets will form, and their energy is probably the trigger that disturbs and compresses molecular clouds sufficiently to continue the star-making process. Supernovae are the self-sustaining force that maintains the birth of stars in galaxies. We will look at galaxies in the next chapter, but before we do we should look at some objects that are not normally considered under the subject of mass loss.

## GLOBULAR CLUSTERS

All the examples of mass loss that we have illustrated so far have been associated with small groups of stars, often single individuals. However, there is plenty of evidence for objects that seem to have both formed their stars and been swept clear of star-forming materials very early in the history of the Galaxy. These are the globular clusters, the fossilised evidence of ancient mass loss on a grand scale.

The first stars to form in the early Universe must have appeared in an environment that was almost devoid of 'metals', elements heavier than hydrogen and helium. At this distant time, over 10 thousand million years ago, there were probably few galaxies of the kind we know now, certainly none as rich in dust as the Milky Way. But there were vast chaotic clouds of hydrogen, in the form of proto-galaxies, poised to make stars.

It seems probable that many of the very massive stars that formed under these conditions appeared quickly in a series of compact 'starbursts' of a kind we no longer see, at least not nearby. As the massive stars turned into red supergiants and exploded as supernovae they soon vanished. If many massive stars had been formed at about the same time and the supernovae explosions had occurred over a similarly short period the effect would have been to sweep their environment clear of gas and dust, halting further star formation, but leaving behind an enormous, compact cluster of stars. The hordes of less massive, longer-lived stars that were also formed in these early starbursts now survive in huge numbers in the ancient, spectacular constellations that we see as globular clusters.

Very few colour pictures have been made of globular clusters, mainly because they are not especially colourful, but they are of enormous interest to astronomers because of their age and their place in the evolutionary history of the Galaxy. Recently, much to everyone's surprise, Galactic globulars have been found to contain pulsars, normally the remnants of young stars that have exploded recently as supernovae. The origin of this enigma is beyond the scope of this book, but the most prolific source of these unexpected pulsars is the globular cluster 47 Tucanae (NGC 104), a new colour picture of which appears in Fig. 7.29. 47 Tuc shows a marked increase in concentration of stars towards its centre, greatly increasing the possibility of the kind of interaction between stars that can produce a pulsar.

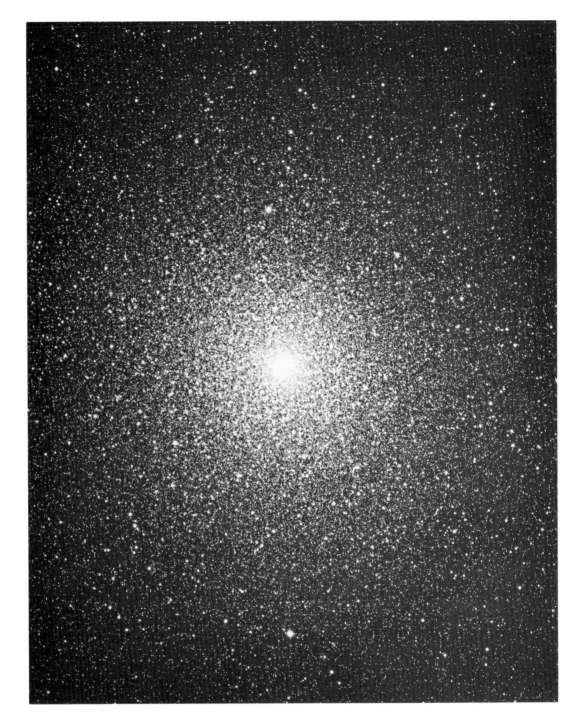

*Fig. 7.29.* The globular cluster 47 Tucanae (NGC 104) was stripped of its star-forming material many billions of years ago, long before the Milky Way formed. The core of the cluster is so crowded that these ancient stars sometimes interact, producing pulsars. Though both blue and yellow stars can be seen on the original, their combined light has a colour similar to that of the Sun, hence the cluster photographs as a pale yellow.

Not all globular clusters are as crowded as 47 Tuc. A short exposure of a more typical example, M5, NGC 5904 (Fig. 7.30), shows that the stars near its centre seem well separated. Closer inspection also shows them to be coloured: the brightest yellow, the more numerous but fainter members distinctly blue. While the faint blue stars far outnumber their yellow companions, the cooler stars are by far the brightest. The integrated colour of the cluster thus appears almost neutral, perhaps

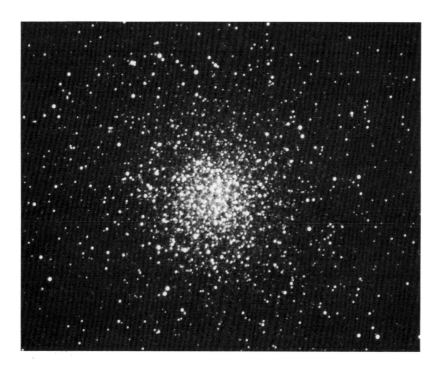

*Fig. 7.30.* The core of the globular cluster M5 (NGC 5904) is a less crowded cluster than 47 Tucanae and its integrated light is somewhat bluer than the Sun.

slightly blue, contrary to the popular belief among astronomers that globular clusters are red.

Although Galactic globulars are ancient and show only very subtle colours, the combined light of stars of different hues, many globular clusters in the LMC are much younger, and in Hodge 11 (Fig. 7.31) the colours of the brightest individual stars can be seen. This cluster is unusual in that many of its brightest stars are blue, and careful measurement of their colours shows that Hodge 11 is a relatively young cluster, which is why its integrated light is clearly blueish.

The Milky Way's family of globular clusters – about 100 are known – evidently did not form in the metal-rich plane of the Galaxy where star formation is currently active. Indeed, it is likely that they formed before the Milky Way *had* a plane, or looked anything like it does now. The orbits of globular clusters carry them far above the flat disk of the Galaxy, probably remembering the distribution of gas in its proto-galaxy phase. Today, their paths send them plunging through the Galactic bulge and the inner parts of the spiral arms.

It was by studying the puzzling distribution and distances of a few of these conspicuous cities of stars that American astronomer Harlow Shapley made a grand leap of the imagination, and suggested in 1918 that they were in orbit around the hidden nucleus of the Milky Way. This was before the Milky Way was understood to be a spiral galaxy and before it was know to have a nucleus. With remarkable insight he thus proposed that the Sun was not in the centre of the Galaxy, or indeed in any special location. The philosophical implications of these observations have been profound, especially since they provided a new sense of scale for galaxies in general and the Universe itself.

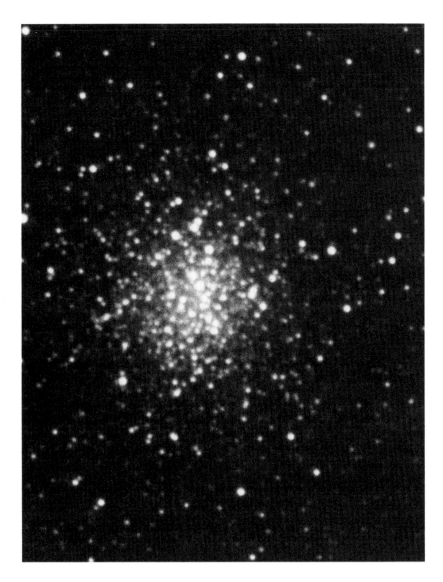

*Fig. 7.31.* Some globular clusters in the LMC are relatively young and contain an unusual number of blue stars. This example is Hodge 11.

While the distribution in space of these enormous, self-gravitating collections of millions of stars led to a new understanding of the size and distance of galaxies, this definition of globular clusters also fits some of the lesser galaxies, and these fascinating objects are discussed in the next chapter.

# 8 GALAXIES

## INTRODUCTION

Galaxies are vast assemblies of stars, gas and dust, so massive that their ingredients have been confined by their own gravity since they were bundled together in the early Universe. These closed but steadily evolving cities of stars vary enormously in mass and may contain between a million ($10^6$) and one thousand billion ($10^{12}$) stars.

Such an amazing range of mass brings with it an equally astonishing range of sizes, shapes and properties, as we will see. The lower end of the scale of mass overlaps with that of the largest globular clusters, which can thus be regarded as miniature galaxies; however, low-mass galaxies are quite distinctive. At the high end of the range of sizes we find enormous individual galaxies with the mass of a whole cluster of lesser examples. Among these giants are some of the most peculiar and extreme objects known.

As in globular clusters, a galaxy is held together by gravity, the mutual attraction of all its constituents. Unrestrained, this force would cause the galaxy to collapse upon itself, but, as with the Sun and its planets, the inward pull of gravity is balanced by the orbital energy of the stars, gas and dust. As with the planets, provided the components keep moving around the common centre of mass they will find its attraction eternally resistable. Since stars in some galaxies are as old as the Universe itself, this strategy seems to work, though black holes at the centre of some galaxies may suggest it is not completely infallible. The pattern of the orbits traversed by the stars is not some academic concept, since the appearance of a galaxy is determined by the distribution of the stars it contains.

It should also be mentioned here that the shapes of galaxies, especially spiral systems, are also determined by the so-called missing or unseen mass, whose presence can only be inferred from the effect that it has on the orbits of the stars. A change in the orbits of the stars induced either by the unseen mass or prompted by some intruding object may change the appearance of a galaxy quite dramatically.

Photographs of galaxies show that they seem to occur in two main sorts. The most eye-catching are the spirals, which have beautiful wound-up arms like catherine wheels, bright features often delineated by clumps of young stars. In nearby examples of spiral galaxies the pictures also show clear evidence of dark clouds of dust and the bright nebulae that are the signs of active star formation, often associated with brilliant knots of stars strung out like irregular shiny beads along the spiral arms.

By contrast, ellipticals have a smooth light distribution and look like fuzzy blobs, which might have a shape anywhere between an elongated American football and a sphere. Whereas spirals are flat and appear dramatically different depending upon their orientation on the sky, ellipticals seem much the same no matter what the viewpoint. In spirals, the stars in the arms are in mainly circular orbits in a flat plane, which is why we see such galaxies edge-on as elongated objects, such as NGC 891 (Fig. 4.11). In ellipticals, the orbits of the hordes of individual stars are much more random, and these galaxies seem to have rounded, fuzzy, rather uninteresting shapes and ill-defined borders. Generally in ellipticals star formation is undetectable and most of them contain very little gas and dust.

To confuse matters somewhat, many gas-rich, star-forming spirals have what seem to be a distinctly 'elliptical' component at their centres; sometimes these are so conspicuous as to be called a 'bulge'. However, these central bulges can also be so small as to be invisible, as in NGC 253. The bulge of our own Galaxy is quite substantial, and though we can see only part of it at visible wavelengths because it is mostly hidden by dust, it still forms the most easily visible part of the Milky Way. The size of these central concentrations of stars, relative to the spiral arms, forms the basis of a system of classification of spiral galaxies. Ellipticals, on the other hand, are classified by their only distinguishing feature, their shape.

Spirals thus present a wide variety of forms to the observer, many of them eye-catchingly beautiful and often endowed with a wondrous symmetry. Such galaxies contain large amounts of star-forming material and most of them reveal by their predominantly blue colours that they are still active in the star-forming business. Perhaps this has encouraged astronomers to spend a disproportionate amount of time studying spiral galaxies. That our own Milky Way Galaxy is also spiral has no doubt contributed to the desire to understand others of a similar kind.

Interest in ellipticals has languished by comparison, partly because they appear less interesting and partly because facts about them are harder to obtain. They contain only old, mostly cool stars, which appear yellow in colour pictures, and comparatively little gas and dust to yield clues about their dynamics. Unlike spirals, ellipticals appear at first sight to have neither internal organisation nor external boundary, and until recently the accepted view of these galaxies was that while they were often big, bright and apparently boundless, they were also uniformly boring. Now, however, evidence is accumulating to show that this generalisation is too simple.

Apart from these two main species of galaxy there are other sub-groups, an in-between type, with in-between colours, the so-called S0s (S-zeros), which are sometimes described as armless spirals, and a miscellaneous mixture of irregulars, which are difficult to categorise. There are also galaxies in pairs and small groups that are clearly interacting, their shapes and symmetry disrupted by their mutual gravity in an endless variety of ways. These mixed-up types comprise a tiny fraction of the total, which is dominated by the various kinds of ellipticals with rather fewer spirals.

Although isolated galaxies are found, they are rare and most seem to congregate in clusters of varying degrees of richness and intimacy, ranging from a few to a few thousand members, with spirals tending to dominate in the smaller groups, and ellipticals in the largest. The Milky Way is a major member of one such small gathering, towards the bottom end of the richness scale. This collection of mostly light-weight galaxies is known, for obvious reasons, as the Local Group. The other major member is M31, the famous Andromeda galaxy, the rest are small fry. The Local Group has about 30 known members, but, as with most things concerned with distance in astronomy, its boundaries are ill-defined and for several minor galaxies membership is controversial. However, the Local Group is a distinct grouping of galaxies, generally thought to be about 7 million light years across, with its centre somewhere between its two dominant galaxies, the Milky Way and M31. We have already described the Milky Way in Chapter 4, so for our first look at galaxies beyond our own we should seek out its nearest neighbours in the Local Group, the Magellanic Clouds.

## THE MAGELLANIC CLOUDS AND THE LOCAL GROUP

The Magellanic Clouds were named after the Spanish-Portugese explorer Ferdinand Magellan by his navigator Ernesto Pigafetta in commemoration of a voyage around the world that Magellan did not live to complete. After the ravages of their arduous journey to the unknown southern seas, through what are now the Straits of Magellan, his sailors must have welcomed the calmer waters of the vast ocean they named the Pacific. There they might have had time to enjoy the unfamiliar southern sky as they sailed to Magellan's untimely death in the Phillipines.

Had Magellan's ships continued westwards at southern latitudes instead of re-crossing the equator, Magellan himself might have survived and they would have discovered New Zealand and Australia in 1521. The history of these countries might then have had a distinctly Iberian flavour. As it was, Australia and New Zealand were not placed firmly on the map until the Englishman Captain Cook, on a voyage with the declared purpose of measuring the distance of the Sun by recording a transit of Venus, navigated their eastern coastlines and claimed them for Britain over 250 years later.

The two galaxies that are the Magellanic Clouds are close enough to be seen with the unaided eye, at least to those of us who live in the southern hemisphere. They look like fragments of the Milky Way, detached in some accident, but in fact the reverse seems to be true. The Magellanic Clouds are satellites of our Galaxy and seem to be in the process of being absorbed by it. Already there have been close encounters. Unseen in photographs is a vast loop of hydrogen gas in which both Clouds are enveloped and which extends into the Milky Way itself. This loop, the Magellanic Stream, has been pulled out from the Clouds by a recent near miss. As well as interacting with the Milky Way, the Magellanic Clouds are interacting with each other and the Small Cloud has been seriously diminished by the affair.

Our first picture, Fig. 8.1, was taken with an ordinary camera perched on the end of the AAT and it shows both Magellanic Clouds much as the eye might see them, faint blobs of light, 25° apart and almost devoid of structure. They are suspended in a part of the sky where there are few bright stars, indeed Sir John Herschel, surveying this part of the sky from South Africa in the 1830s, described it as a veritable stellar desert. There is little here to say that these two galaxies and the Milky Way are in any way associated.

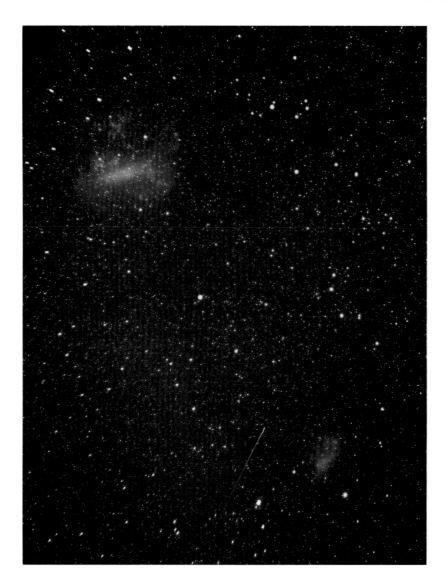

*Fig. 8.1.* The Magellanic Clouds are a pair of interacting galaxies that are so close to the Milky Way that they appear about 25° apart on the sky. They are visible to the unaided eye from any dark southern site and are never set for observers in the southern half of Australia. This wide-angle picture was made on colour film with an ordinary camera.

However, on the wider scale of things, galactic encounters are common and many galaxies seem to have been involved with others at some time. But both Magellanic Clouds are representatives of a fairly unusual kind of galaxy, the so-called gas-rich irregulars. Despite their rarity, there is another example of a gas-rich irregular undergoing a gravitational interaction for which we have a colour picture. This is NGC 4027, seen in Fig. 8.2. Like the LMC, NGC 4027 also has a fainter companion with which it is probably interacting, producing the curved, elongated arm, an analogue of the northern loop that we see in the LMC. Fortunately, because the LMC is so close we can examine its internal arrangement in much more detail than in NGC 4027, which is 30 million light years distant, almost 200 times as far away. These two pictures make an interesting comparison between a 30 millimetre and a 4 metre telescope recording similar kinds of objects.

The LMC is about one tenth of the mass of the Milky Way, while its diminutive companion, the SMC is about a quarter of the size of the larger. This makes the LMC

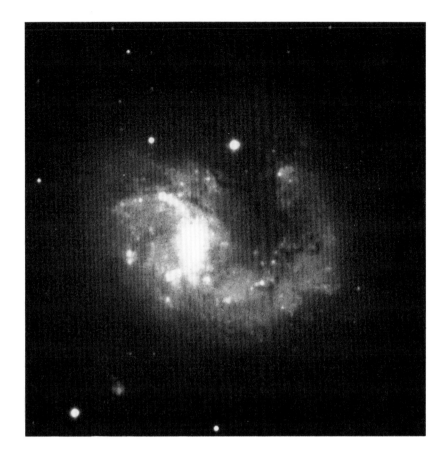

*Fig. 8.2.* Galaxies like the Magellanic Clouds are quite rare, but NGC 4027 is another, much more distant example. Like the LMC it has a smaller companion (not shown) with which it is interacting and which has distorted it. Unlike the Magellanic Clouds, NGC 4027 has no massive Milky Way-like spiral nearby to complicate the entanglement.

quite a substantial galaxy, with a mass equal to about 10 thousand million stars like the Sun. (Galaxies are conventionally weighed in terms of solar masses. The use of this term does not mean that Sun-like stars predominate, and this mass includes gas and dust.) However, because it is so close, the LMC covers a large area of sky, and conventional telescopes have too narrow a field of view to do it justice, which is why Fig. 8.3 was taken on conventional colour film with a normal Hasselblad camera, effectively a telescope with a 30 millimetre aperture.

I made the picture by combining three similar shots of the LMC made with ordinary colour slide film. The process enhances faint detail and contrast but does nothing to improve resolution of fine detail. It shows the large-scale structure of the nearest galaxy to the Milky Way, 170 000 light years away, and the veil of Milky Way stars through which we see it. The predominant colour is blue, because the LMC is rich in young stars. It also shows that the LMC is an asymmetrical galaxy with bright loops (especially to the north) and patches of star formation surrounding an elongated 'bar', around which the optical image of the galaxy is centred. These distortions are the first subtle evidence that the LMC is an interacting system whose equilibrium has been disturbed by encounters with other galaxies.

Fortunately, the proximity of the Magellanic Clouds enables us to study them in great detail, especially their stellar populations. This is as important to astronomers as the study of individual people in different cultures is to sociologists, and for similar reasons. The

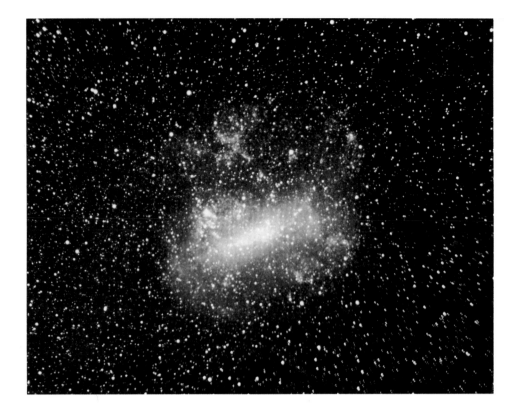

*Fig. 8.3.* This is another image of the LMC and is the result of combining three photographs like that in Fig. 8.1 to increase the contrast. It is thus comparable to the photograph of the distant Magellanic system NGC 4027 seen in Fig. 8.2; however, because the LMC is nearby, many more Galactic stars are visible along the line of sight.

Magellanic Clouds are sufficiently unlike the Milky Way for the star-forming environment to be different, producing an interestingly different stellar population, the more luminous members of which are well within the reach of Earth-bound telescopes. By comparing groups of stars of similar ages in the two galaxies, astronomers can address the same kind of nature–nurture evolutionary problems that affect isolated populations of all kinds.

The UK Schmidt picture reproduced as Fig. 8.4 shows some of the brightest stars and the vivid red star-forming regions. We also see bright blue clumps of stars and numerous patches of dust silhouetted against the luminous background of the galaxy. The largest single bright structure in the LMC is the diffuse 'bar' running diagonally across the picture. It is superimposed on a fainter, yellower ellipsoidal component, which is difficult to see and contains the oldest stars in the galaxy. The whitish stars in the bar suggest that it is relatively young.

Younger still are the irregular clouds of blue stars at the extreme ends of the bar, and younger yet again are the even bluer star clouds whose light dominates north of the galaxy, seen in a separate photograph (Fig. 8.5). Here there are still many active star-forming regions, flagged by bright red nebulae, some excited by extremely massive and luminous stars, which, as we have seen elsewhere, both illuminate the nebulae around them and, in the process, rapidly disperse them. In the LMC many nebulae show the effect of stellar winds very clearly, by blowing bubbles on all kinds of scales.

Fig. 8.6 shows a typical example of a nebula where cavities have been created in the gas and dust by a few very luminous stars. Others are seen in Figs. 7.7–7.9. Such features seem more common in low-mass galaxies such as the Magellanic Clouds than in hefty

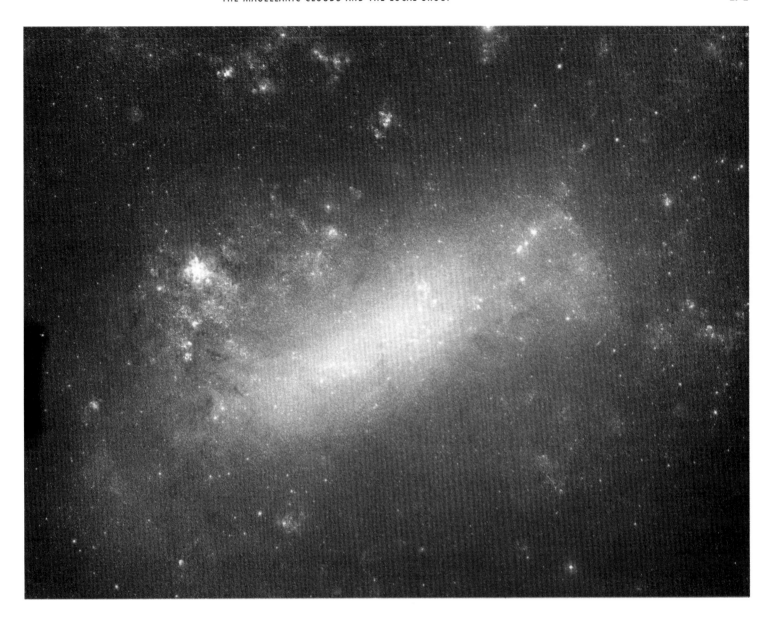

*Fig. 8.4.* The LMC is over 4° across and the brighter parts fit neatly into the 6.5 × 6.5 degree field of the UK Schmidt Telescope. At its eastern (left) end is the enormous star-forming region 30 Doradus, the Tarantula nebula.

spirals such as the Milky Way, possibly because the interstellar material is at the same time both more constrained and more turbulent in our Galaxy, and these delicate structures are swept away before they become recognisable.

Much more disturbed is the spectacular nebulosity that engulfs the eastern end of the bar of the LMC. The brightest part of this is variously known as the 30 Doradus or Tarantula nebula and is visible to the unaided eye, like the Orion nebula in our own Galaxy. It also is of similar angular size, about 1° across, but is 100 times more distant! The nebula in Fig. 8.7 surrounds one of the most vigorous and massive star-forming regions we know anywhere, far exceeding the scale of any we can find in the Milky Way. Many of the thousands of stars we see in this picture are as luminous as the brightest we see with the unaided eye in Orion. If this enormous star-forming region were in the Milky Way, at the distance of the Orion nebula, the night sky would be ablaze with tens of thousands of brilliant stars.

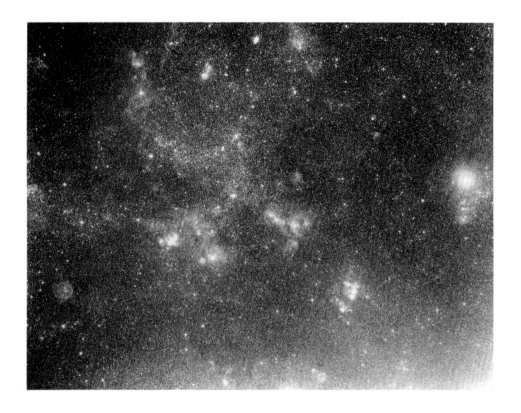

*Fig. 8.5.* The LMC has undergone several recent bouts of star formation, leaving behind clouds of stars dominated by massive blue stars. This enormous constellation is to the north of the galaxy and still contains a few bright star-forming nebulae.

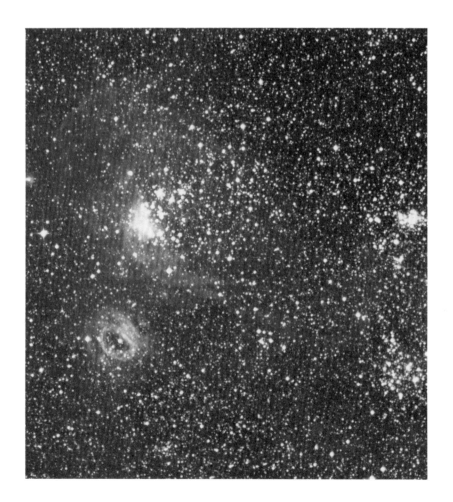

*Fig. 8.6.* Henize 55, a bubble nebula in the LMC.

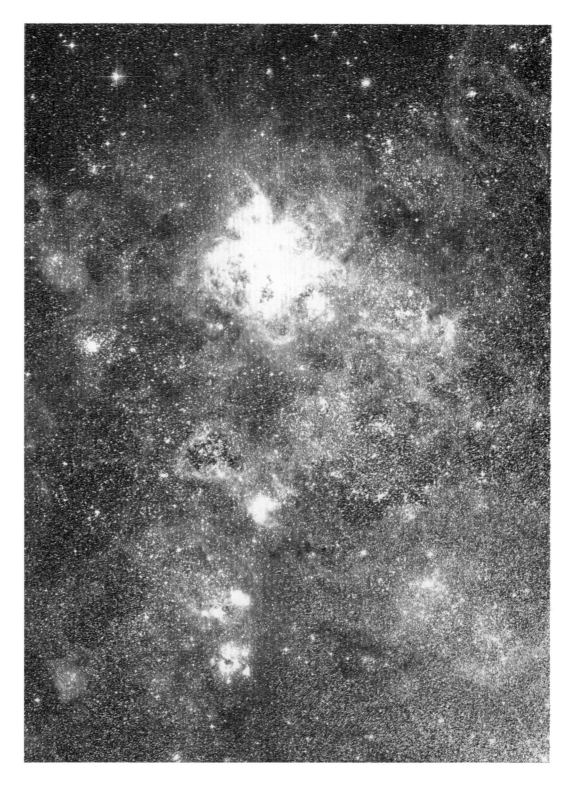

*Fig. 8.7.* At the eastern end of the LMC is one of the most active nearby regions of star formation and it has spawned vast numbers of massive stars. The biggest and brightest nebulosity is the Tarantula nebula seen in more detail in Chapter 6.

In all galaxies, star formation seems to occur sporadically wherever the ingredients are available in the right state and are subject to an appropriate trigger. The trigger may be the result of buffeting from within by supernovae, or of external gravitational encounters with other galaxies. Both these mechanisms seem to be active in the eastern end of

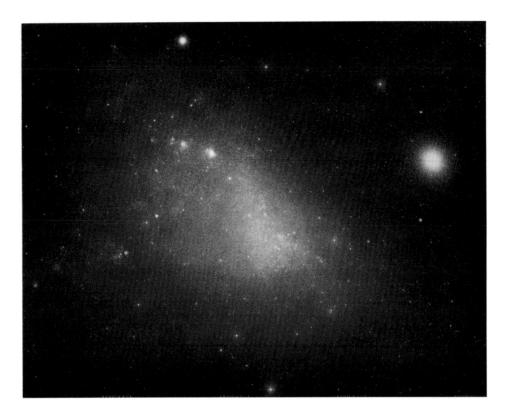

*Fig. 8.8.* The SMC is more distant than the LMC and has fewer bright stars and star-forming nebulae. It is a galaxy that seems to be in the process of disintegration, torn apart by both the LMC and the Milky Way. In the same field of view is the large Galactic globular cluster 47 Tucanae.

the LMC. Radio observations show that the visible part of the 30 Doradus nebula is at the edge of a region where there is a steep increase in the density of hydrogen, probably swept up and compressed by the LMC's interaction with the Milky Way. The star formation this has caused has in turn produced many supernovae, which have themselves prompted the formation of yet more stars. This phenomenon is known as a starburst, and it is fortuitous that such a fine example is here on our doorstep, near enough to be interesting, not too close to be threatening.

In contrast to the vigorous star formation in the LMC, the SMC seems to be a quiescent galaxy, but this tranquil appearance is misleading. Fig. 8.8 shows most of the galaxy and reveals that it has few star-forming regions scattered into no particular pattern. The form of the SMC is indescribable and so it is the perfect example of an irregular galaxy. It also contains fewer hot, young stars than its neighbour the LMC. This is curious, since it is now known that the SMC has undergone a much greater degree of disruption as a result of its gravitational encounters than has its larger companion. Why, then is there not the violent burst of star formation that seems to accompany such events?

The answer probably lies in the mass of the SMC, which is too low to retain the hydrogen it once held against the purging effect of close encounters with the Galaxy and the LMC. The SMC has been stripped of the means of making more stars, its hydrogen dispersed beyond recall into the Magellanic Stream. This gas will eventually fall into the LMC and the Milky Way, creating localised starbursts as it does so. In the meantime the SMC has joined the ranks of those many galaxies whose ability to regenerate has been destroyed by straying too close to more massive companions.

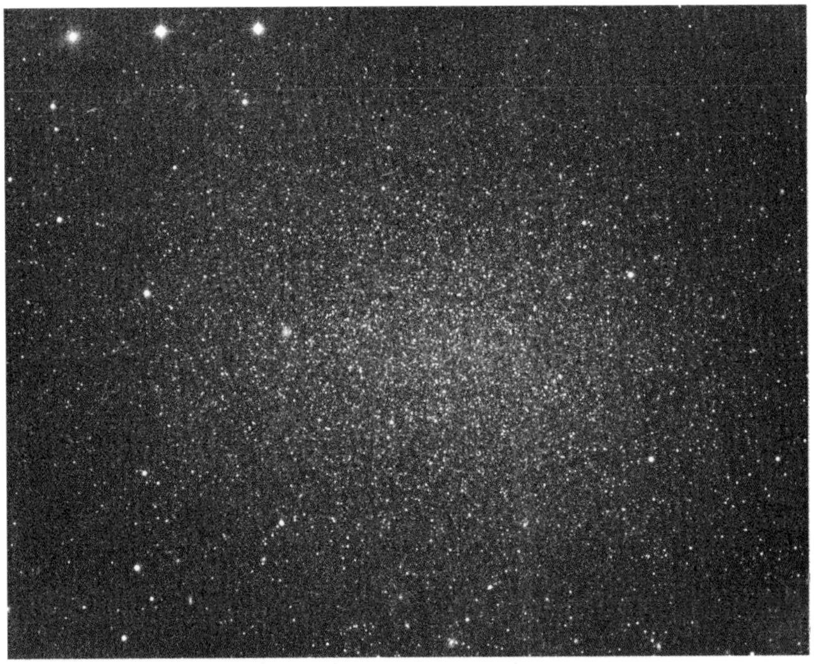

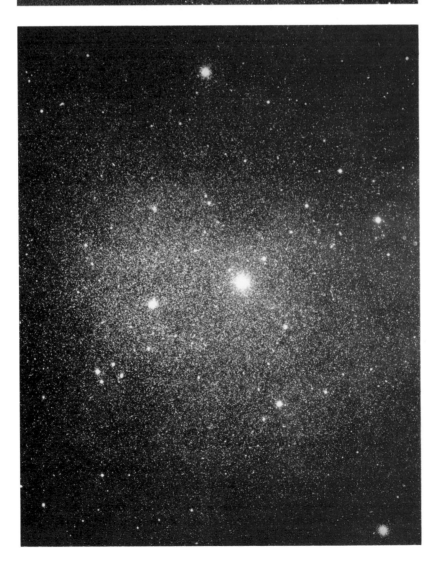

*Fig. 8.9.* The Leo I dwarf elliptical galaxy is one of eight or so low-mass galaxies that are gravitationally bound to the Milky Way.

*Fig. 8.10.* The Milky Way's retinue of dwarf companions remained undiscovered until photography was used to find objects too faint to be seen by eye. This example is the Fornax dwarf galaxy, discovered in 1977.

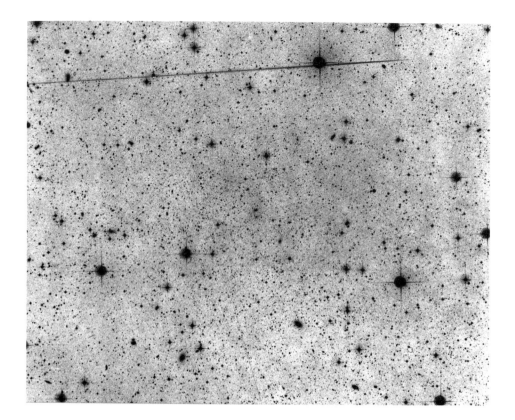

*Fig. 8.11.* This is the most recent dwarf galaxy discovery, the Sextans dwarf, discovered on UK Schmidt plates by a measuring machine in Cambridge, England, in 1990. The galaxy is just visible on this negative print as a slight increase in the number of faint stars towards the centre of the picture.

Not only is the SMC almost devoid of the means to make more stars, it seems to be in the process of disintegration. Careful observations to determine the distance of stars within the galaxy strongly suggest that it is enormously elongated along our line of sight, so much so that it may be fragmenting into several miniature galaxies. These groups of stars will probably lose their identity as components of a galaxy as the interaction continues, and may eventually merge with the LMC and the Milky Way.

The massive Milky Way, equivalent to 1000 billion stars like the Sun, is able to retain satellites other than these highly visible clouds of stars, but most of them are so faint that photography had to be invented before they could be discovered. For obvious reasons these are known as dwarf galaxies, and among the most distant of these insignificant collections of stars is one known as Leo I. It is so identified because a second dwarf (Leo II!) was discovered in the same constellation after the first was found.

Leo I and II are extremely light-weight by galactic standards and are about 700 000 light years away, the most distant of the small colony of nine or so galaxies that are bound to the Milky Way. Despite their distance both are resolved into stars by a big telescope. Very few photographs of them exist because they represent a considerable photographic challenge and Fig. 8.9 is certainly the only colour photograph of a resolved dwarf galaxy to have been taken. The Leo I dwarf is a particular problem because this faint galaxy appears very close to the brilliant star Regulus whose scattered light overwhelms it.

Another well-studied Milky Way companion is in the southern constellation of Fornax and is known, with admirable consistency, as the Fornax dwarf. At a distance of

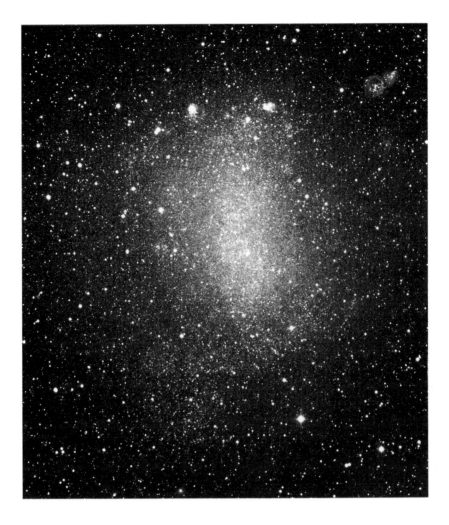

*Fig. 8.12.* NGC 6822 is a member of the Local Group of galaxies. Though it appears relatively undisturbed, there is evidence of relatively recent star formation in the scattering of bright stars across the face of the galaxy.

450 000 light years it is closer to us than those in Leo and is well resolved into stars. The Fornax dwarf even boasts a few tiny globular clusters among its stellar population (Fig. 8.10). At the time of writing, the final entry in this family of nine or so faint galaxies was being discovered by a measuring machine in Cambridge, England, which was scanning a plate taken with the UK Schmidt Telescope in Australia. Our photograph of the Sextans dwarf, Fig. 8.11, is reproduced as a negative and is the result of photographically combining the deepest images obtained on three plates. Despite this, the galaxy is hard to spot against the foreground Milky Way stars, though it fills most of the photograph. The Sextans dwarf is almost 300 000 light years away.

The galaxies that are bound to the Milky Way are more than mere near-neighbours. The orbits that they traverse enable us to calculate our Galaxy's mass, and, very surprisingly, it turns out to be much greater than can be accounted for by the known, visible ingredients. By studying these relatively insignificant, inert collections of stars we have begun to realise that the cause of most of the gravitational might of the Milky Way and other galaxies like it cannot be seen, or even detected, at any wavelength or by any technique we can bring to bear. This realisation is naturally quite alarming, especially since we have no idea if this

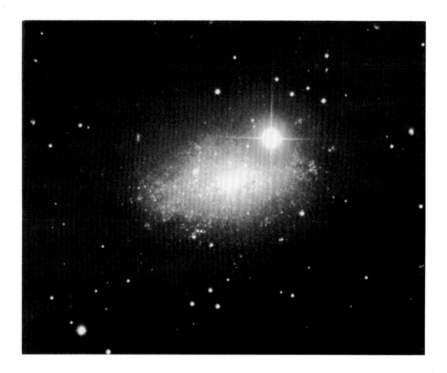

*Fig. 8.13.* IC 5152 is a small galaxy of predominantly blue stars, probably beyond the Local Group, perhaps 5 million light years away. The bright star is in the Milky Way, at most a few hundred light years distant.

'missing mass' is due to strange atomic particles, clapped-out stars or even vast numbers of Jupiter-sized planetary bodies.

Beyond the cluster of galaxies that make up the Milky Way's family, ten times more distant than the LMC, is another Local Group member with a mass similar to the disrupted SMC. Although it is nearly 2 million light years away, NGC 6822 is still close enough to be resolved into individual stars with the AAT, and three-colour photography (Fig. 8.12) reveals their distribution clearly. At one end of a prominent bar a few clouds of glowing red gas can be seen. As in the LMC, some of these are the wind-blown bubbles so typical of low mass galaxies (see Fig. 7.9). At the other end of the bar, bright blueish stars straggle out into what appears to be the beginnings of a spiral arm. The irregular distribution of the scattered bright blue stars is in marked contrast to the underlying smooth ellipsoid of large numbers of much fainter yellow stars that are the galaxy's old population.

Although clearly a member of the Local Group, NGC 6822 is relatively isolated and it appears to be undisturbed, looking perhaps much as did the SMC before its disruption. Despite its relative isolation, the galaxy has relatively recently undergone a burst of star formation that produced the sprinkling of blue stars seen on Fig. 8.12. It is not obvious what happened to turn the star formation on, and it is equally difficult to find out what turned it off, or at least slowed it down, since there seems to be hardly any star formation occurring now in comparison to the explosive activity in the LMC. Ample hydrogen remains, however; the galaxy sits inside a huge cloud of it, enough to make 100 million stars like the Sun.

In Fig. 8.13, IC 5152 is also resolved into stars, which means that it is relatively nearby. However, despite its proximity, this small galaxy is probably just beyond the Local Group, which is loosely defined as the collection of 30 or so galaxies within 2 million light years of the Milky Way. Apart from the Milky Way and the similarly

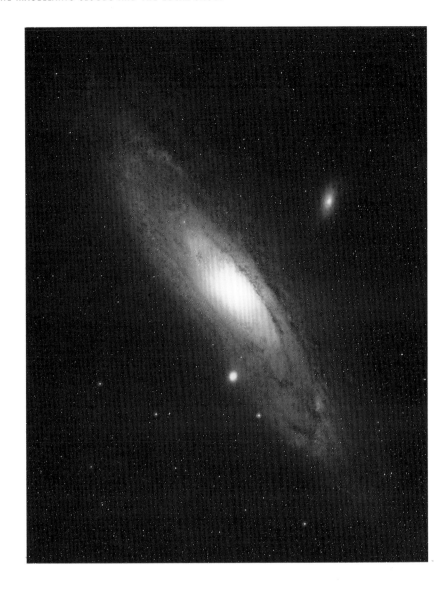

*Fig. 8.14.* M31 (NGC 224) is a galaxy like the Milky Way in size and shape, about 2 million light years away. Like the Milky Way it has many fainter companion galaxies, the brightest of which are elliptical galaxies, unlike the gas-rich Magellanic Clouds. This picture was made with the Palomar 48 inch (now the Oschin 1.2 metre) Schmidt telescope in California.

massive M31 galaxy in Andromeda, most of our immediate extra-galactic neighbours are relatively light-weight collections of stars and gas like this. The bright object that appears at the eastern end of the galaxy is an eighth magnitude blue star in the Milky Way.

By far the most imposing member of the Local Group is the spiral system M31 (NGC 224), *the* galaxy in Andromeda. Apart from the Magellanic clouds this is the only other galaxy visible to the unaided eye and is much the most distant object we can see without optical aid. I am unable to provide a three-colour image of this spectacular galaxy in its entirety because it barely skims the horizon of the Anglo-Australian Observatory's telescopes, but some years ago I remade a duplicate of the colour film image of M31 for Mt Wilson and Las Campanas Observatory. The original was on colour reversal (slide) film and was taken with the Palomar Schmidt Telescope by (the then) Palomar Observatory's photographer Bill Miller. It is reproduced as Fig. 8.14.

Like the Milky Way, M31 has attracted some lesser companions. Among them is NGC

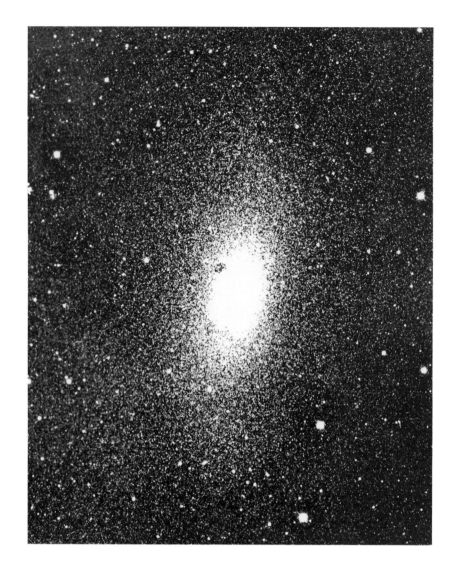

*Fig. 8.15.* One of M31's companions is NGC 205 which is quite close enough to be resolved into stars on this excellent Kitt Peak 4 metre telescope photograph. It has been specially printed to show the dust clouds in the interior of the galaxy.

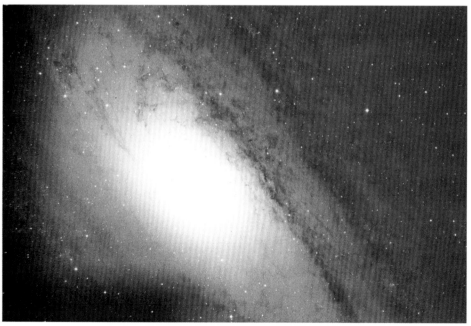

*Fig. 8.16.* The central bulge of M31 is composed mostly of billions of old, faint, yellowish stars. As in our own Galaxy, streaks of dust hide some of it from view. Photograph from the Isaac Newton Telescope.

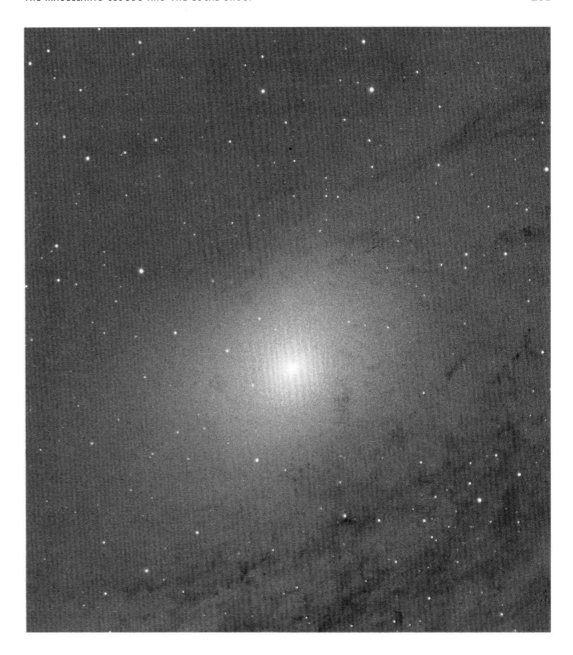

*Fig. 8.17.* At the heart of M31 is a tiny nucleus, which appears slightly blue and somewhat elongated on this picture, which was made from the same plates as Fig. 8.16.

205, one of the few elliptical galaxies close enough to be resolved into stars with an Earth-bound telescope. It is shown in Fig. 8.15, a picture made from a plate taken with the Kitt Peak Observatory's 4 metre telescope, which I copied during a visit to the Kitt Peak plate archives in Tucson. The purpose of the visit was to explore the plates for the most distant galaxies, but nearer examples such as this could not be ignored. Unsharp masking revealed patches of dust towards the nucleus; such distinct dustiness is unusual in elliptical galaxies.

The central bulge of M31 is very substantial (Fig. 8.16) and against its almost uniform brightness can be seen the innermost dust lanes, which are typical of a spiral galaxy and very reminiscent of those seen in the Milky Way. I made this picture during a visit to the Isaac Newton Telescope on the island of La Palma in 1991. Unlike the Milky Way,

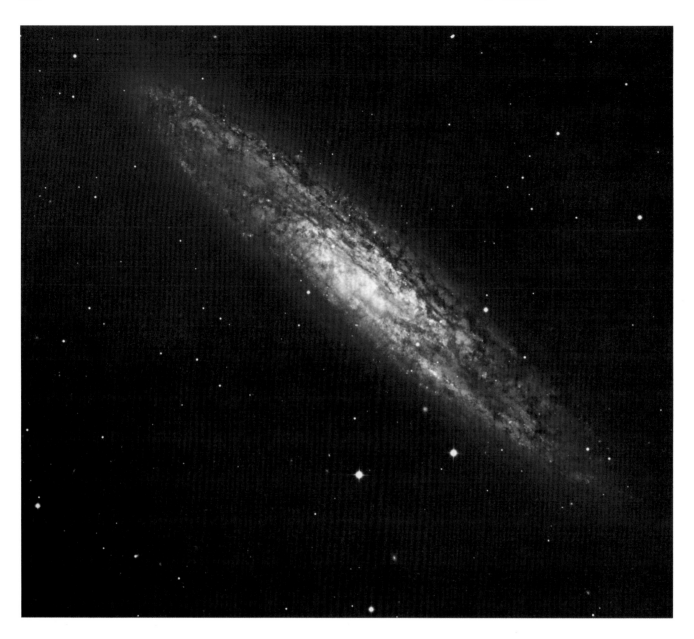

however, we can see the nucleus of M31 and it turns out to be slightly blue, very small, bright and elongated. It is seen in Fig. 8.17, a picture made from the same plates as Fig. 8.16, but here I used unsharp masking to explore the structure in the brightest parts of the nearest large Milky Way-like galaxy.

## BEYOND THE LOCAL GROUP

The proximity of the Local Group galaxies ensures that we have had good pictures of most of them to display and enjoy for many years, but it was not until large telescopes were built in the southern hemisphere that the beautiful spirals of the next nearest group

*Fig. 8.18.* NGC 253 is seen from below its northern edge, revealing subtle signs of spiral arms that are almost hidden by the galaxy's dust and inclination. This galaxy is the most massive of the sparse Sculptor Group which straddles the south galactic pole.

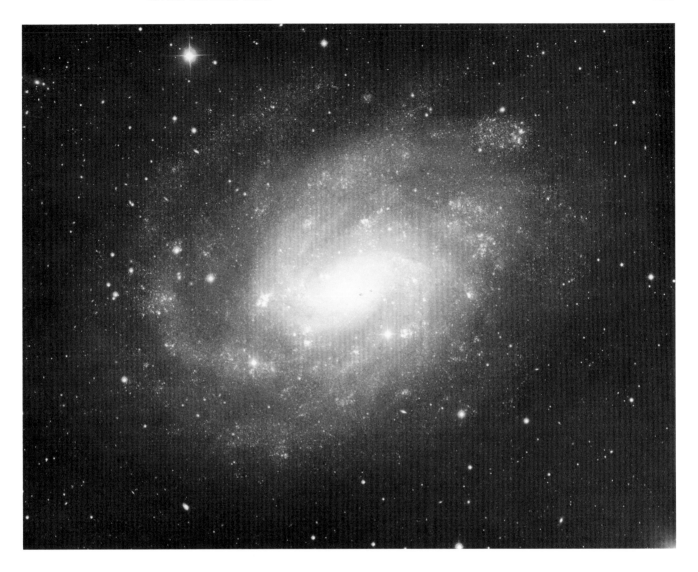

Fig. 8.19. NGC 300 is a galaxy of quite low surface brightness but large angular size. Most of the light comes from the young population in the delicate spiral arms and thus appears blue on colour photographs.

of galaxies could be properly studied. These galaxies straggle the southern constellation of Sculptor, which also happens to contain the south galactic pole. In this direction we look out of the Milky Way perpendicular to the plane of the Galaxy and so there are relatively few nearby stars and hardly any dust to complicate our view of the more distant Universe.

The Sculptor Group is dominated by spiral galaxies, the most substantial of which is NGC 253. If this were a northern hemisphere object visible from the latitude of Paris it would no doubt have had a Messier number, and probably a common name referring to its elongated form, perhaps Le Saucisson. As it is, this bright edge-on spiral is comparatively little known, despite being only 10 million light years distant. The galaxy is seen almost edge-on and Fig. 8.18 shows us a perspective that reveals a dusty rim in silhouette against the clouds of stars behind it. This in turn tells us that the top right (north-west) part of the galaxy is the nearest to us. NGC 253 is inclined almost 80° to our line of sight, which is why it seems so elongated.

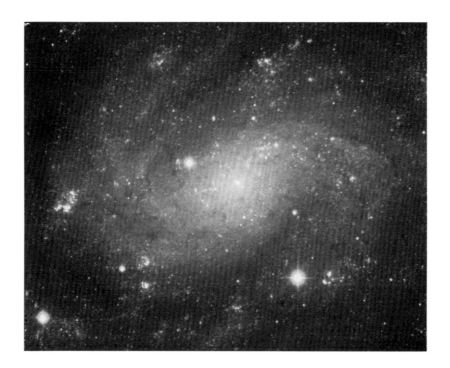

*Fig. 8.20.* The inner regions of NGC 300 are dominated by clumps of young stars and nebulae, but a faint haze of older stars can be seen around the star-like nucleus. During the green-light exposure the atmospheric 'seeing' worsened briefly, which is why the brightest stars have unnatural green haloes.

While we see NGC 253 almost edge-on, any observers on a planet surrounding a star in NGC 253 would see the Milky Way as an enormous spiral, exactly face-on, much as we see M83, but much closer. This is because NGC 253 is located in the direction of the pole of rotation of the Milky Way, as the Milky Way is in the direction of M83's pole. In just the same way that we can define the direction of the Earth's axis of rotation in the sky we can also locate the axis of rotation of our Galaxy.

This position in the sky is at right angles to the richly starred plane of the Milky Way. In this direction, we look out to the extra-galactic Universe through the thinnest part of our Galaxy, which is why there are so few foreground stars in images of galaxies in the Sculptor Group. However, an inhabitant of NGC 253 would be lucky to see much at all since it is one of the dustiest known galaxies, and, because of its inclination, most of its internal structure is hidden from us. This is another perspective on the currently controversial topic of the opacity or otherwise of spiral galaxies. NGC 253 is a particularly opaque example. The plates for this picture were exposed by Ken Freeman during the commissioning period of the AAT before I joined the Anglo-Australian Observatory.

In contrast to NGC 253, which is somewhat smaller than the Milky Way, NGC 300 is petite, a delicate, transparent spiral. This galaxy has a relatively low mass and, quite unlike NGC 253, it contains relatively little dust. Distant galaxies can readily be seen through its spiral arms. Fig. 8.19 is a recent colour picture of this beautiful galaxy, and shows the disk to be blue, as expected from a system containing so many young stars. However, the galaxy is intrinsically quite faint, so the photographic plates that recorded the image were copied at higher than normal contrast to make Fig. 8.19, thus losing some detail in the brighter central regions.

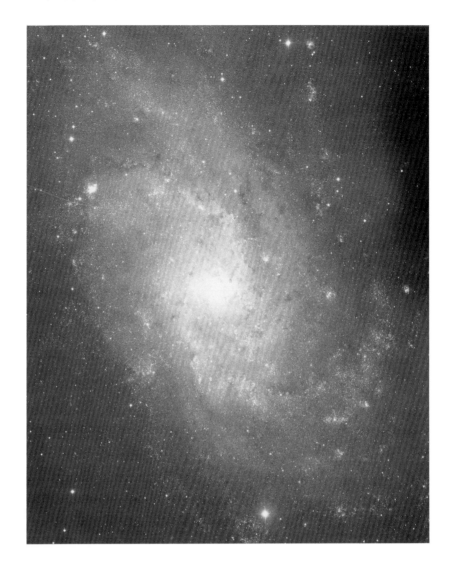

*Fig. 8.21.* M33 (NGC 598) is the spiral galaxy in Triangulum, and is the northern hemisphere's counterpart to NGC 300. It, too, is large but faint, and two pictures are needed to do it justice. This photograph, made from plates taken with the Isaac Newton Telescope on La Palma, shows the faintest outer parts.

I had wanted to make a colour picture of this galaxy ever since I saw how well it was resolved into stars on a *V* (green light) plate I made for Don Mathewson in 1980. A *B* (blue light) plate was obtained for another, unrelated project by Ray Sharples in 1988, and I finally completed the three-colour set with a red light exposure in telescope time awarded specifically for three-colour photography in 1991. Even now, the less than perfect *V* plate produces unnatural green haloes around the brightest stars, but the pressure on telescope time means that the plate is unlikely to be repeated. Involvement in astronomy certainly leads to an appreciation of the fullness of time.

A second image of NGC 300, which I made from the same plates but with a lower contrast, shows its nucleus as a tiny, bright, star-like object, visible in Fig. 8.20. It is surrounded by a haze of faint, unresolved stars, too few in number to be called a nuclear bulge, but that is what it is. Traces of dust can be seen as dark patches silhouetted against this background and it is here that the inner spiral arms seem to originate. Although still a member of the Sculptor Group, we can so clearly see individual stars in NGC 300 that it

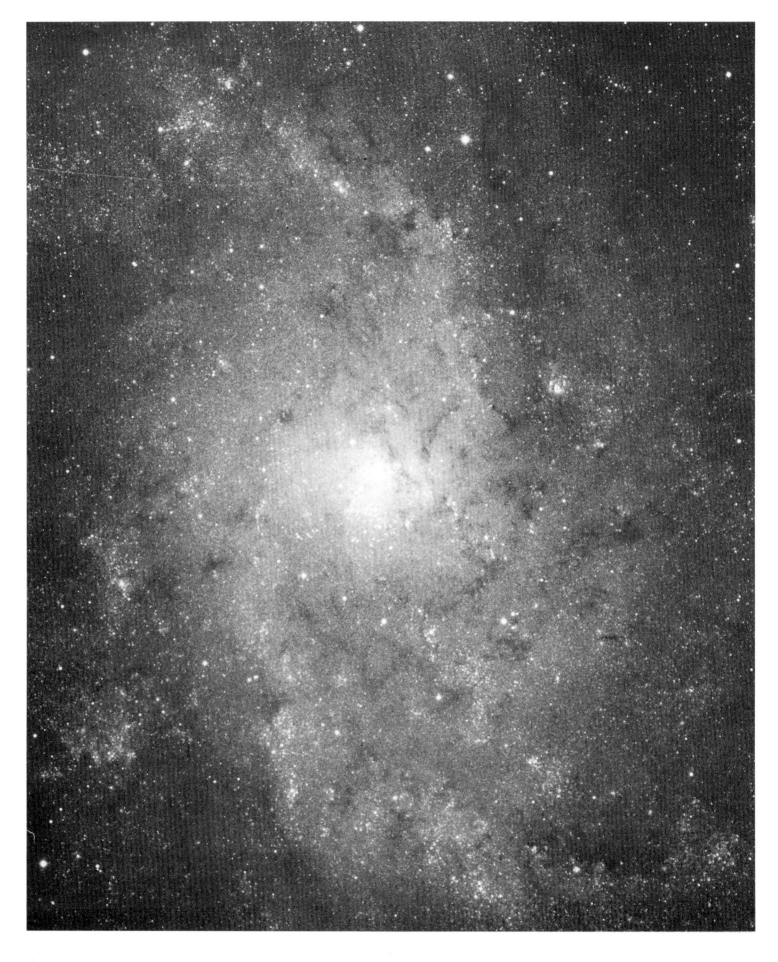

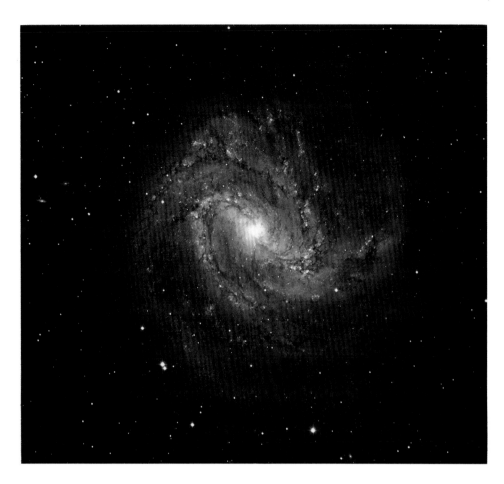

*Fig. 8.23.* M83, NGC 5236, one of many fine galaxies in the Centaurus–Hydra region.

*Fig. 8.22.* The inner regions of M33 show that, like NGC 300, it has a stellar nucleus at the centre of a haze of faint stars, but many more dusty clouds, as befits a more massive galaxy.

appears to be closer to the Milky Way than NGC 253. Recent estimates put the distance of NGC 300 at almost 7 million light years.

Again from a recent visit to the Isaac Newton Telescope I include for comparison (Figs 8.21 and 8.22) pictures of M33 (NGC 598), the galaxy in Triangulum, NGC 300's northern near-equivalent. It, too, has a stellar nucleus, but the galaxy is somewhat more luminous and has more and brighter star-forming regions. In both galaxies they are well scattered and seem to be disconnected from the few dust patches. M33 has about the same mass as the LMC but is much more distant, about 2.8 million light years away.

I offer no excuse for duplicating as Fig. 8.23 the picture of M83 that was used in Chapter 4 as a model for the appearance of the Milky Way, except to say that it is one of the most magnificent galaxies in the sky. This object is in Hydra, closer to the Galactic plane than the Sculptor Group galaxies, hence the greater number of foreground stars. Even so, plenty of detail is visible and a few of the most luminous stars in M83 can be seen as individuals, despite its distance, about 10 million light years.

Much more massive than either NGC 300 or M33, M83's active ingredients seem to be organised somewhat differently. The pink star-forming nebulae in the galaxy are neatly lined up along the inner edges of its spiral arms, as is the dust from which they spring. Between the arms are relatively fainter regions, seemingly devoid of stars, and the arms themselves are bright and well defined. None of these observations is true of NGC

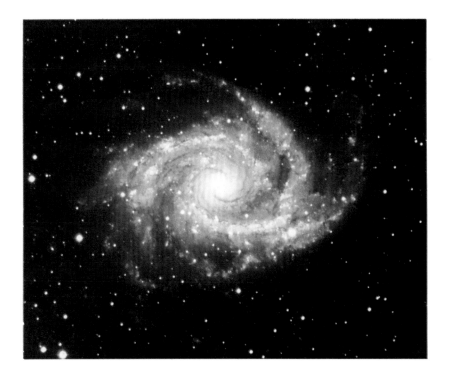

*Fig. 8.24.* One of the most beautiful spirals, NGC 2997 is among the most popular pictures produced by the Anglo-Australian Observatory. Its blue arms and yellow nucleus perfectly illustrate the basic types of star that are found in spiral systems.

300, underlining the enormous variety of often subtle differences between these complex but fascinating objects. On the other hand, in both galaxies the blue stars seem to occur in clumps, suggesting that star formation in the spiral arms of a galaxy is in some way infectious.

At a distance of about 35 million light years, the brightest stars of NGC 2997 are unresolved by Earth-bound telescopes, though our plates were taken under less than ideal conditions. Like M83, it is a magnificent spiral system that clearly reveals its internal organisation by its colour. Fig. 8.24 shows blue spiral arms, peppered with clumps of young stars and the pinkish nebulae from which they came. The arms seem to arise in the yellow nucleus as dust lanes which spiral outwards a considerable distance before fattening out, inflated by newly-formed stars. Spiral galaxies are quite flat, and NGC 2997 is inclined at about 45° to our line of sight and so appears ellipsoidal rather than round.

Elongation is only partly responsible for the unusual appearance of NGC 1365 (Fig. 8.25). This beautifully symmetrical galaxy in Fornax is a barred spiral, one of the finest examples in the sky, and galaxies with a bar as pronounced as this are quite rare. NGC 1365 is also a massive galaxy, perhaps as substantial as the Milky Way, which may itself be a less extreme type of barred galaxy, though it is hard to tell from within. In the other spirals we have seen, the central bulge of yellow stars seems to have no visible connection to the arms that emerge from it. In NGC 1365, the yellow stars *are* the bar and seem to lead directly into the curved arms where the most of the star-forming activity is.

Seen from above its pole, NGC 1566 (Fig. 8.26) does not seem so different from many other spiral galaxies scattered across the sky, though it is more symmetrical than most. Even the colour differentiation between its outer and inner parts is muted. But this

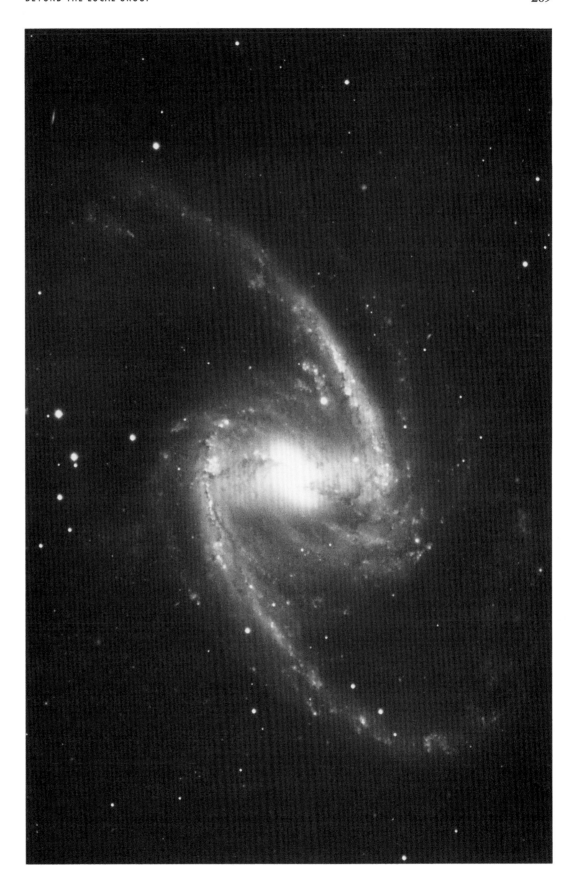

*Fig. 8.25.* One of the most striking barred spiral galaxies is found in the southern constellation of Fornax. NGC 1365 is 60 million light years away.

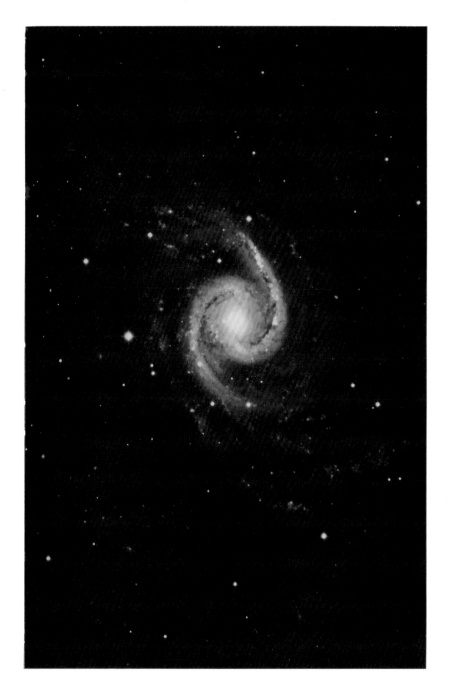

*Fig. 8.26.* The tightly wound, remarkably symmetrical spiral arms of NGC 1566 show very little colour differentiation. Even its active nucleus has an unexpected blueish tinge.

is a Seyfert galaxy, a spiral system in which there is an unusual amount of activity in the nucleus. NGC 1566 is the brightest member of a nearby group in the southern constellation of Dorado and is at a distance of about 50 million light years.

Apart from its beautifully symmetrical, tightly-wound spiral arms, which are shown (in Fig. 8.27) to extend much further than in the colour picture, the galaxy is of interest because its very luminous nucleus has many of the characteristics of a quasar, a galaxy so bright (and usually so distant) that it appears star-like, even with a powerful telescope. NGC 1566 is much less energetic than a quasar, but has some

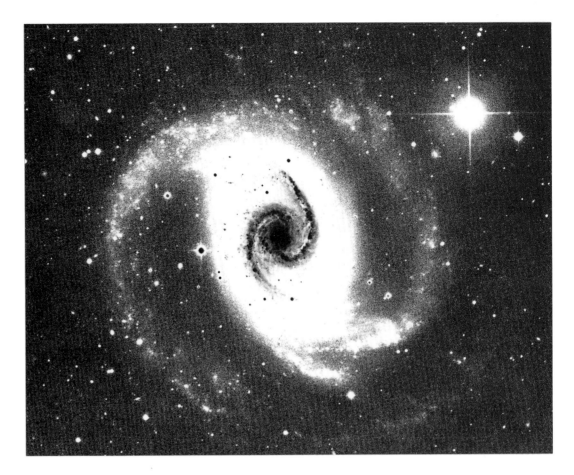

*Fig. 8.27.* The colour picture shows only part of NGC 1566. By adding together all three plates used to make the colour picture in Fig. 8.26, the enormous extent of the spiral arms becomes visible. In this picture the negative and positive images are printed to the same scale and in register.

features in common. Galaxies of this type are named after Carl Seyfert, the American astronomer who first recognised their peculiar nature in 1943. The active region at the centre of NGC 1566 has recently been found to vary on a time-scale of less than a month, which indicates that it is extremely compact for a galactic nucleus. Spectra show that hot gas near the tiny nucleus is moving with an abnormally high velocity, which hints that it may be in orbit around a massive black hole at the heart of NGC 1566.

While NGC 1566 has the outward appearance of a normal galaxy, indeed its symmetry is remarkable, it has just enough intriguing features for some astronomers to describe it as peculiar, which brings us to another large category of extra-galactic objects.

## PECULIAR GALAXIES

Like people, most galaxies are peculiar in some way or another and there are very few, if any, that can be described as completely normal. In both people and galaxies oddities are numerous, and, like galaxies, even apparently normal people can turn out to be strange on better acquaintance. To add further confusion to this problem of definition in galaxies, normality depends strongly on how one looks at the galaxy, more particularly

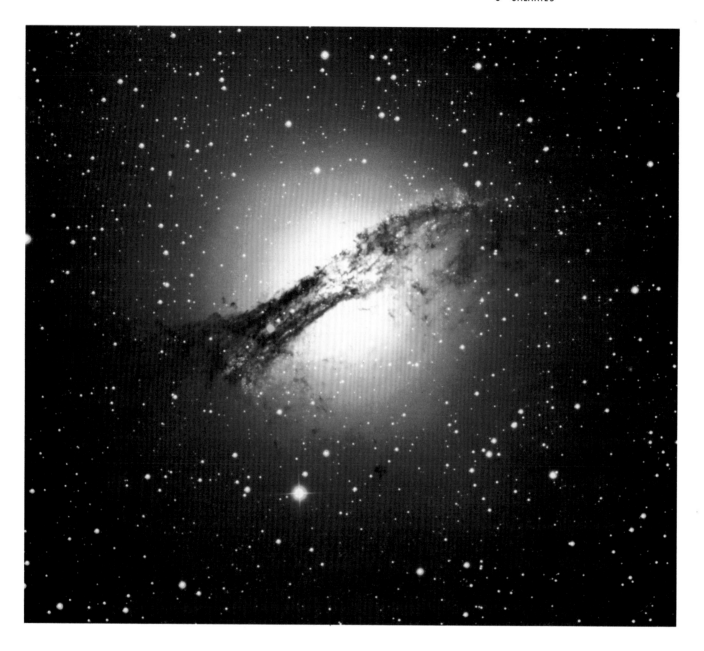

on how faint its outer parts are. Some photographic processes, such as unsharp masking or photographic amplification, are specifically intended to uncover their peculiar properties in much the same way that a doctor might use new techniques to seek out a hidden disorder in a seemingly healthy patient. Even if a photograph of a galaxy looks normal at first sight the chances are high that a thorough examination will find something odd about it.

It is also true to say that we know much less about the fundamentals of galaxies than we do about people. We are still not sure what triggers their formation, how they evolve and why there are different kinds, nor do we know in detail why they congregate as they do and how they die. It is therefore presumptuous to claim an intimate knowledge of any of them, including the one in which we live. But our ignorance of what

*Fig. 8.28.* The famous southern radio source Centaurus A (NGC 5128), the nearest strong radio galaxy. The colour picture shows the effect of dust absorption on the colour of starlight in the dust lane and clumps of blue stars at either end of it.

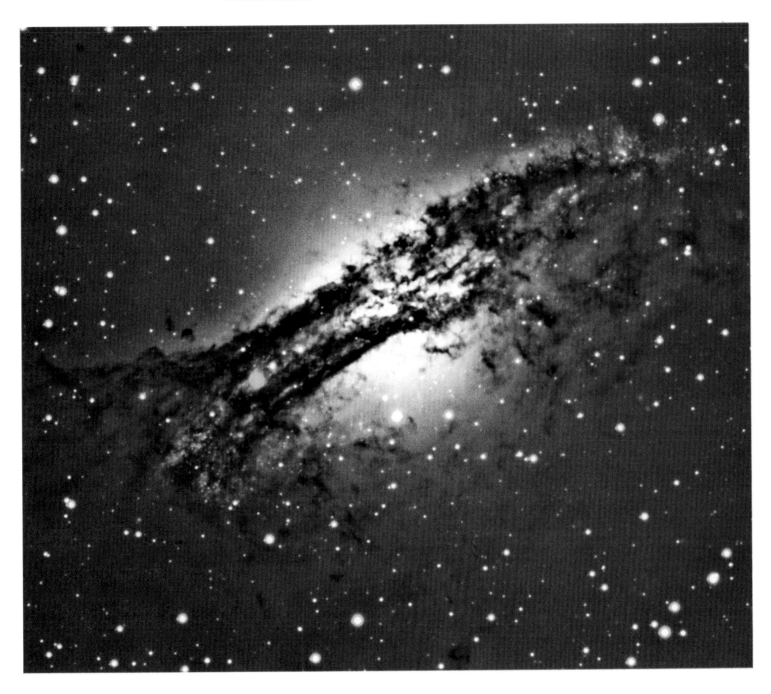

Fig. 8.29. Within the dust lane of Centaurus A are numerous small, pink nebulae which are the sites of star formation. The blue-green 'star' here is a supernova that was bright when two of the three plates for the colour picture were taken.

makes people odd does not stop us talking about them, indeed it makes them more interesting, and so it is with galaxies.

The first example we choose is a famous southern radio source, Centaurus A (NGC 5128), an object with strong Australian connections. This galaxy had long been labelled as peculiar, or at least noteworthy, since it was first sketched by James Dunlop at Governor Brisbane's Paramatta Observatory in the 1820s. The slightly irregular, finely structured dark band that seems to divide it into two is very unusual, and until little more than 30 years ago it was not known if the object was a nearby component of our

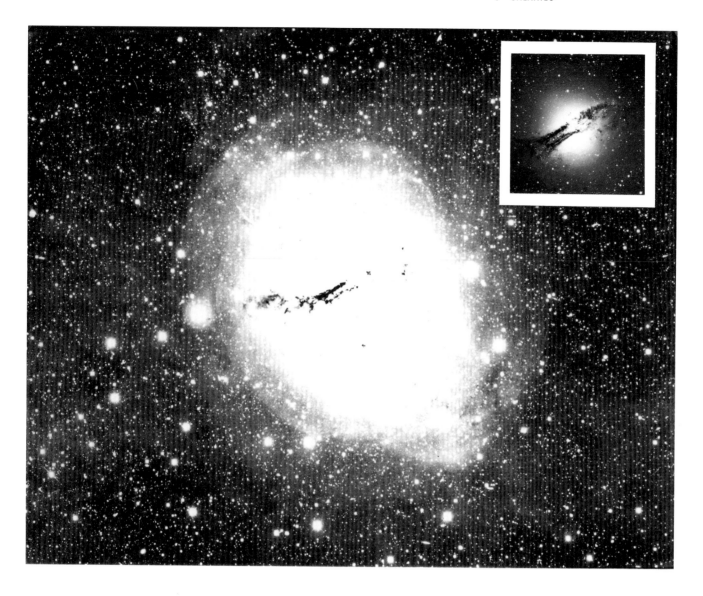

Galaxy, perhaps even a peculiar planetary nebula (Chapter 7) or a much more distant object far beyond the Milky Way.

We now know that it is at least 10 million light years away and that its diffuse, more or less circular 'body' consists of thousands of millions of old stars, typical of many giant elliptical galaxies. But Centaurus A is not a typical galaxy. Fig. 8.28 shows that the dark lane is a huge dusty curtain which obscures, dims and reddens the light of stars behind it. At the ends of the dust lane small blue patches of younger stars can be seen, together with a few pinkish star-forming regions. Fig. 8.28 is an excellent example of the extra information provided by a colour image.

The name 'Centaurus A' (or, informally, 'Cen A') indicates that this most unusual object is the brightest radio source in that constellation. It was one of the first radio sources to be positively identified with an optical object, by pioneer Australian radio astronomers in the 1950s. Apart from radio waves, the galaxy is a copious

*Fig. 8.30.* Careful image manipulation of long exposure plates of Centaurus A reveal a series of subtle inner structures (shells) that strongly hint that the galaxy we see now has been disturbed by absorbing at least one other in the last billion years. Inset is the 'normal' appearance of the galaxy, derived from the same plate, which was taken on the Cerro Tololo 4 metre telescope by John Graham.

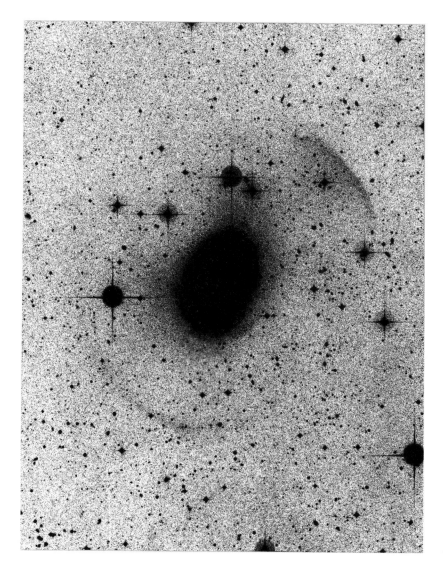

*Fig. 8.31.* Many elliptical galaxies show evidence of shell-like structures if their photographic images are enhanced by special techniques. The faint outer shells of NGC 1344 shown in this negative print were among the first to be discovered.

source of X-rays and gamma rays as well as infrared and visible radiation. The prominent dust lane and powerful radio radiation are believed to be the remnants of a dusty spiral, which fell into the more massive giant elliptical galaxy a few billion years ago.

The green- and blue-light plates for this picture were taken for a project unconnected with colour photography on different nights in May 1986. Quite by chance, this was shortly after a supernova had been discovered in the galaxy, again by an Australian, the Reverend Robert Evans, who lives near Sydney. Unfortunately, cloud prevented me from completing the three-colour set on a subsequent night and the red-light plate had to wait until the following year, by which time Evans' supernova had faded, which is why one star in the dust lane appears to be an unexpected blue-green (i.e. minus red) colour. This is more clearly seen in Fig. 8.29, which I made from the same three plates as Fig. 8.28, but using a special copying technique that emphasises the small structures while eliminating most of the featureless image of the elliptical galaxy itself. This reveals

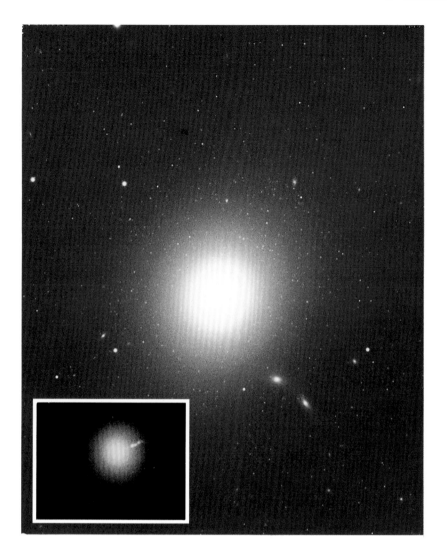

*Fig. 8.32.* The giant elliptical galaxy M87 in Virgo looks smooth and featureless and its colour suggests its light is dominated by old, cool stars. However, at its core is a massive black hole from which emerges an energetic jet, seen inset.

much more detail in the dust lane, especially the compact pink nebulae and the clusters of blue stars they have produced.

The application of photographic amplification and unsharp masking to another plate of Centaurus A, taken in Chile by Australian astronomer John Graham, wrings out another, quite unexpected peculiarity. I was in Chile at the invitation of the Cerro Tololo and ESO Observatories to discuss photography with their staff. John offered this plate as a suitable subject for demonstrating unsharp masking and we very quickly discovered that the faint outer envelope of the galaxy embraces a series of incomplete arcs or shells, illustrated in Fig. 8.30. These features are of very low contrast and are further evidence of the arrival of an interloper.

From the distribution and size of the shells, Peter Quinn was able to calculate that Centaurus A has encountered another, less massive galaxy, perhaps something the size of M33. That merger is still far from consummated and the intense activity that makes Cen A such a strong radio source is only the latest manifestation of an interaction that has been under way for about a billion years. The interaction also led to a very

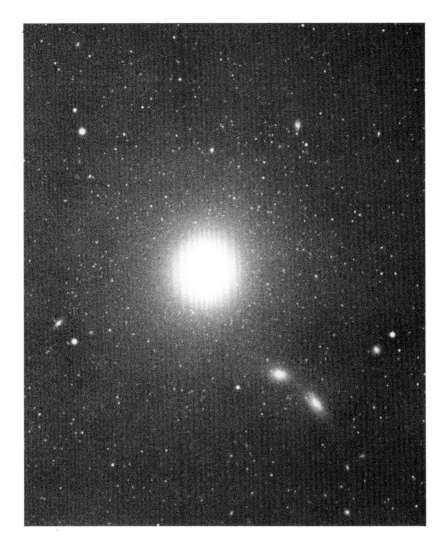

*Fig. 8.33.* Like many massive galaxies, including the Milky Way, M87 is surrounded by a cloud of globular clusters, revealed by an unsharp masking process. Unfortunately, the masking process has upset the colour balance of the picture near the two smaller galaxies.

nice paper by Malin, Quinn and Graham on the dynamics of this fascinating southern galaxy.

The improved photographic emulsions of the last two decades and the photographic techniques they have triggered have shown that a surprisingly large number of otherwise normal-looking ellipticals had shell-like structures of the kind found in Cen A. Among the first to be discovered were those around NGC 1344, a prominent, unusually isolated elliptical galaxy. I first noticed the brightest shell simply by looking at a film copy of the UK Schmidt Telescope's southern sky survey, which showed a faint but extensive arc, apparently unconnected to the galaxy, but clearly centred on it. The staff at the UK Schmidt Telescope willingly made available another original plate for photographic amplification that confirmed the reality of the brighter shell and revealed another, fainter one.

Too faint to be recorded in colour, I show as a negative in Fig. 8.31 observational evidence of what now seems to be a common occurrence, the non-violent, almost orderly merging of galaxies that leaves behind long-lived traces of the event in the form of deli-

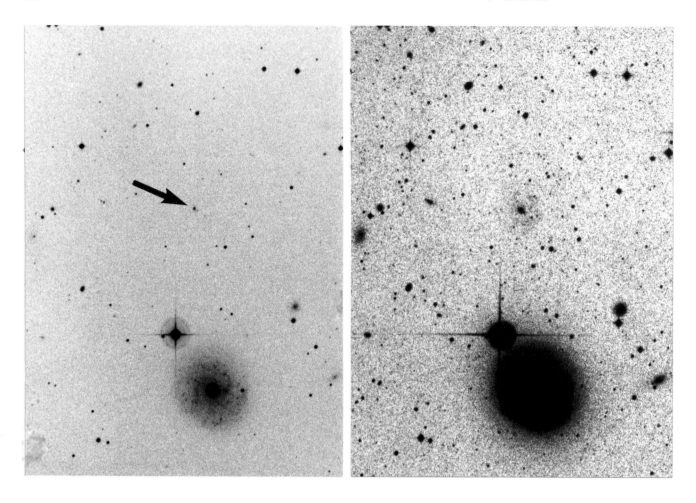

cate shells. The merger hypothesis came from Peter Quinn and, with the help of Dave Carter, it was not long before inspection of the southern sky survey turned up over 140 examples of galaxies with faint outer shells.

Elliptical galaxies do not have to have shells to make them peculiar. NGC 4486 (M87) in Virgo appears quite normal in Fig. 8.32, but, as the strongest radio source in Virgo (i.e. Virgo A), it clearly is not. This is another galaxy where there are powerful signs of a massive black hole in its centre. The evidence this time comes in the form of a curious jet emanating from the nucleus (inset).

M87 is surrounded by hordes of globular clusters, which are seen in Fig. 8.33 as a clustering of faint star-like images centred on the galaxy. This gigantic system is the most massive member of a nearby rich cluster of galaxies in the constellation of Virgo, which has been the subject of numerous studies, especially directed towards establishing the composition of the cluster's galaxy population. As a consequence of one of those studies with colleagues Greg Bothun and Chris Impey, both of whom work in the USA, I found yet another kind of peculiar galaxy that my co-workers dubbed Malin-1 in the paper announcing its discovery.

Malin-1, seen arrowed in Fig. 8.34, is the least interesting-looking of all the peculiar galaxies in this chapter, yet it is in some ways the most extreme. Though it was discov-

*Fig. 8.34.* This insignificant-looking galaxy (Malin-1, arrowed) is among the most massive known, but is very inefficient at producing stars, which is why it is faint. The combination of relatively few stars and large size ensure that such objects are hard to find. The bright galaxy in the foreground is the Virgo cluster spiral NGC 4571. The arrowed image (left) was made by simply copying a UK Schmidt plate. The right-hand photograph combines four separate, photographically amplified plates. Even so, Malin-1 remains little more than a smudge.

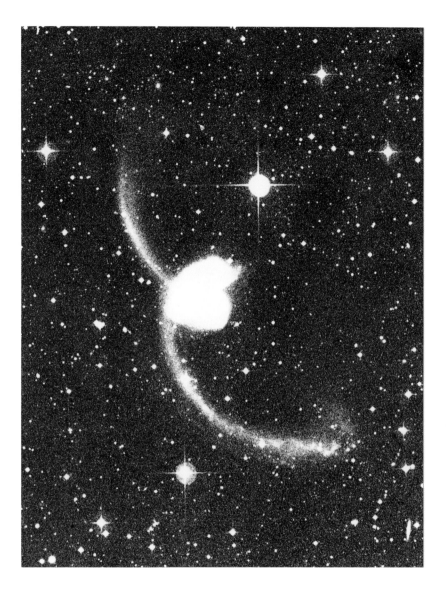

*Fig. 8.35.* The Antennae (NGC 4038–39) are two colliding spiral galaxies which have produced enormous faint tails composed of stars displaced during the interaction.

ered while I was working on plates of the Virgo cluster, it is far beyond those galaxies, over 100 times more distant. Its remoteness is indicated by its velocity of recession, 28 000 kilometres per second, whereas on average the Virgo cluster galaxies move away from us (or us from them) at a more modest 1200 kilometres or so per second. Though the optical image is unexciting, observations of the hydrogen content at radio wavelengths show that Malin-1 is a uniquely massive object, containing enough gas to make 1000 galaxies like the Milky Way. The puzzle is why is it not 1000 times as luminous as the Milky Way.

It seems that this huge, crouching, almost invisible giant has just started to make stars. Around the slightly extended nucleus that first drew my attention to it are the merest traces of spiral arms, containing a few star-forming regions. A galaxy as rich in hydrogen as this should appear as bright as (though smaller than) the foreground Virgo galaxy NGC 4571 in the lower part of Fig. 8.34. Malin-1 is probably a proto-galaxy. Perhaps 15 thousand million

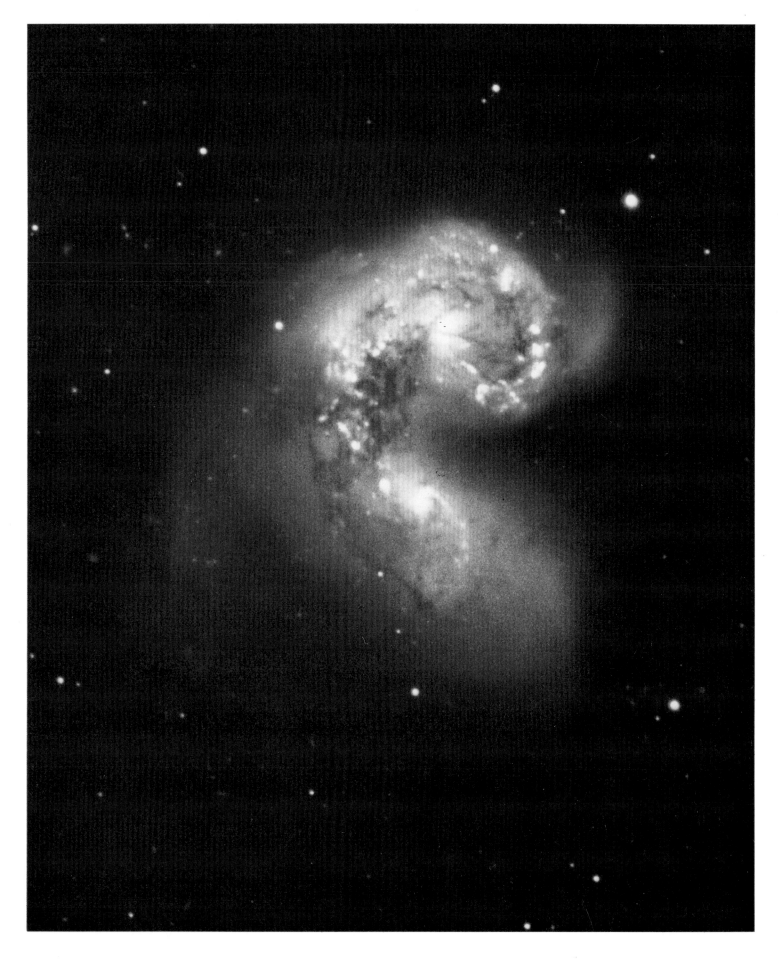

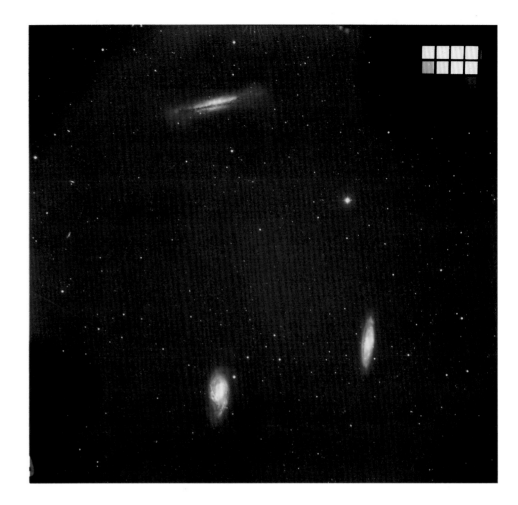

*Fig. 8.37.* The interacting group of galaxies in Leo includes M65 and M66 and the edge-on system NGC 3628. This trio of spirals fits neatly into the 1° field of the AAT's prime focus.

*Fig. 8.36.* A closer look at NGC 4038–39, the galaxies involved in Fig. 8.35, shows the detail of their interaction and the disruption of their internal structure. Clouds of recently formed blue stars provide most of the light, against which dust is seen as dark yellow-brown. The two galactic nuclei are still visible as diffuse yellow blobs.

years after most galaxies started to make stars, here was the first example of a galaxy that failed to ignite its hydrogen in the form of stars soon after the Big Bang. There are probably many more galaxies of this type, possibly even fainter, and my colleagues encouraged me to search for them by making it number one. Unfortunately, most such galaxies will be much fainter than the night sky air-glow, and are therefore very difficult to find unless they have distinguishing features such as a distinctive nucleus.

Prominent among the galaxies classified as peculiar are those involved in interactions with others. We have already seen how a recent encounter has encouraged star formation in the LMC and almost destroyed its smaller companion, and how ancient mergers leave their signatures after 1000 million years. Our next example is a demonstration of an extreme interaction that is still under way. Looking like some giant cosmic insect, the curiously curved arms of the Antennae galaxy demand an explanation. It was forthcoming in 1973, in a paper by the Toomre brothers, neither of whom was an astronomer.

They considered the now well-known image of these two galaxies (Fig. 8.35, NGC 4038–39) and devised a model that explains much of what is seen. Two spirals approached each other, their disks rotating in opposite directions. As they interacted the energy stored in their rotation was sufficient to project two plumes of stars in curved

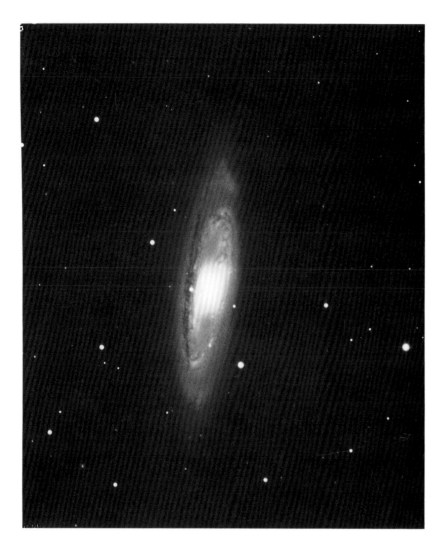

*Fig. 8.38.* M65 (NGC 3623) appears to be a fairly normal spiral, but it is involved in an interaction with its near-neighbours, though it is itself apparently little altered by the encounter.

paths in more or less opposite directions. These plumes are seen in the negative image of Fig. 8.35. A good deal of rotational energy (angular momentum) was lost in this spectacular display, but that is not the end of the affair for these galaxies.

If the galaxies had been ellipticals, where the stars are in much less ordered orbits, there would have been much less angular momentum to dissipate and long spindly arms would have been an unlikely outcome. But not only were these galaxies spirals, at least one of them was of the gas-rich variety, and the effect of a gravitational interaction on gas is to provoke it into making stars, as we saw in the LMC.

Paul Murdin, David Allen and I showed a picture of the Antennae in our book *Catalogue of the Universe*, which was published in 1979. It was clear from the UK Schmidt plates I used for that photograph that the brighter parts of galaxies involved contained a wealth of fine detail, barely resolved with the UK Schmidt Telescope, but the chance to re-photograph them with the better plate scale of the AAT did not occur until 1991. Now, a new AAT colour picture of the merging pair, Fig. 8.36, shows a wealth of detail.

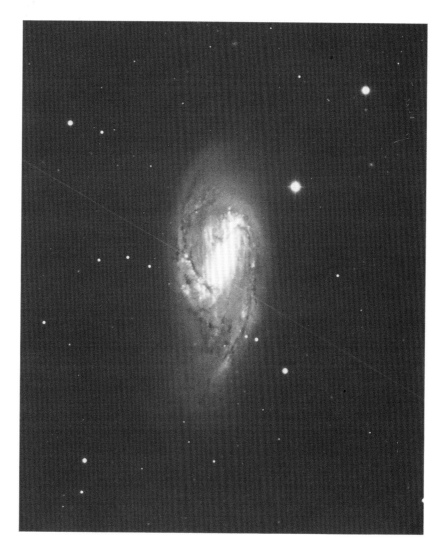

Fig. 8.39. M66 (NGC 3627) has several anomalies which identify it as the prime mover in the interaction with NGC 3628 (Fig. 8.40).

Much of the information here is new, because this is the first true-colour picture of this galactic get-together. Not only can we differentiate the clumpy clouds of new stars by their colour, and the distribution of ribbons of dust by its effect on the colour of stars behind it, but the remaining, relatively undisturbed parts of the galaxies in the lower part of Fig. 8.36 are also subtly different colours, which suggests that one of the galaxies was an S0 rather than normal spiral. S0s are disk galaxies relatively free from gas and whose colour is on average yellower than that of the typical gassy catherine wheel. This would explain why most of the starburst activity is in only one of the pair of galaxies: that is where the gas is.

From the colour picture we can also identify as separate entities the still relatively unscathed nuclei of the two galaxies. They are seen as two extended yellowish regions, one in the heat of the action, the other below, in what remains of what must once have been a substantial S0 galaxy. If and when the nuclei merge there will no doubt be more celestial fireworks, but there will still be plenty to see for millions of years to come.

Not all interactions between galaxies are as obvious as the Antennae. In the constel-

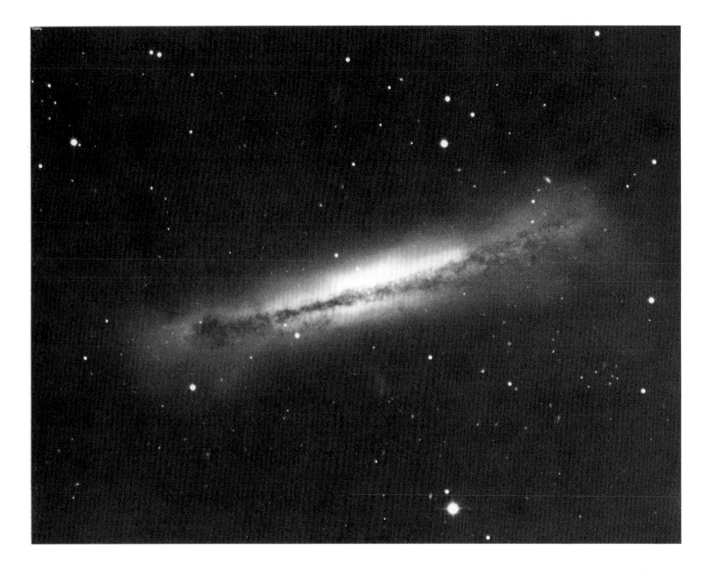

lation of Leo there is a small, compact group of galaxies known as the Leo triplet, comprising M65 (NGC 3623), M66 (NGC 3627) and a galaxy too faint to attract Messier's attention, NGC 3628. We see from Fig. 8.37 that with careful positioning of the guide probe and the calibrator, the three of them fit neatly into the 1° square field of the AAT. The apparent proximity of these galaxies is not a line-of-sight coincidence. These galaxies are together in space and their closeness has affected each of them in different ways.

A look at the individual members of the group reveals that each of them is in some way disturbed. M65 (Fig. 8.38) has a peculiar arm, while M66, which is a beautiful spiral with a well-developed central bulge (Fig. 8.39), also has some peculiarities, in particular the sharp kink part way along the most obvious arm. The most interesting of the three galaxies is the one that Messier missed and is seen in Fig. 8.40. NGC 3628 is one of the few good examples of a galaxy that is by chance oriented edge-on to us, and it is the only one that I am aware of that is in an interacting group.

The effect of the interaction on NGC 3628 is clear evidence of the delicacy of the

*Fig. 8.40.* The almost edge-on spiral galaxy NGC 3628 shows subtle signs of disturbance. The dust lane is slightly askew and the fragile ends of the spiral arms have been fluffed up by interaction with its companion galaxies.

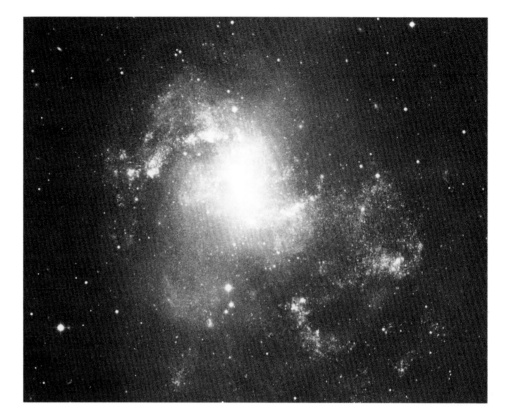

*Fig. 8.41.* NGC 1313 is a relatively nearby southern hemisphere starburst galaxy. The conspicuous clouds of bright stars left behind by sporadic star-forming activity give little indication that the main, invisible mass of the system is largely undisturbed.

disks of spiral galaxies. Perturbations from its distant companions have puffed out the ends of the arms and set the dust lane askew in a way that might be very difficult to see if the galaxy had been differently oriented. The central bulge of the galaxy is seen sticking out above and below the disk and it too shows signs of disturbance in its peculiar 'boxy' shape. This may be the remains of an earlier encounter when the galaxy was young, evidence preserved in the orbits of the stars around the nucleus. A galaxy of the same kind and with a similar, edge-on orientation is NGC 891, seen in Fig. 4.11.

If the interacting galaxies of the Leo triplet could be separated as one might isolate disruptive children, NGC 3628 and its companions might regain their equilibrium and symmetry, but this is unlikely to happen. The galaxies are bound together by gravity. After a few more close encounters they may lose their individuality and other, more damaging interactions may occur. Eventually the remains might become a massive elliptical galaxy.

Some galaxies show the signs of interaction, asymmetrical structures and erratic star formation, without any evidence of invasion by another galaxy. Just why this is so is not known, and such 'starburst' objects are currently a 'hot topic' in the study of galaxies. One example for which we have a colour photograph is NGC 1313. Its appearance in visible light is shown in Fig. 8.41, but recent radio observations reveal that the gas in this galaxy behaves as though it is a perfectly ordinary, well-behaved spiral, though the irregular star-formation pattern seen here suggests otherwise. Through combined optical–radio observations of this kind it is becoming apparent that sporadic star formation can occur without any recognisable trigger such as an encounter with another

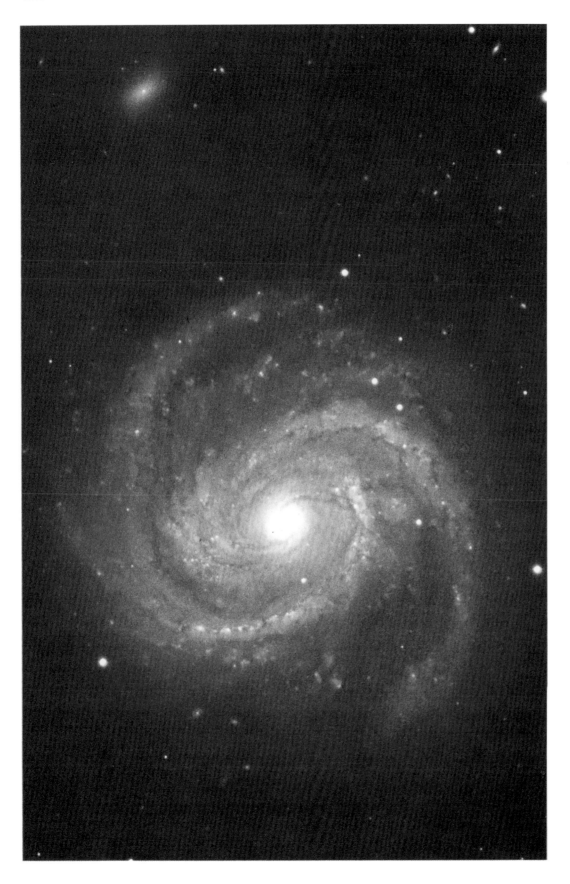

*Fig. 8.42.* M100 (NGC 4321) is one of the most spectacular spiral galaxies in the nearby Virgo cluster. At the top of the picture is one of the commonest types of galaxy in rich clusters, a dwarf elliptical, this one with a bright nucleus.

galaxy, and, once started, it breaks out apparently at random, producing patches of blue stars scattered across the face of the galaxy.

We end this look at the endlessly fascinating interactions between galaxies with another glance at the Virgo cluster, a rich gathering of galaxies where close encounters are very likely. We begin by examining a colour picture of M100, NGC 4321, one of the finest spiral galaxies in the group. In Fig. 8.42 it appears to be a fairly normal example, though some asymmetry is seen, with the southern half of the galaxy apparently producing stars at a greater rate than the northern part. In the same photograph is a small elliptical galaxy with what seems to be a bright star at its centre. This is a nucleated dwarf elliptical, and there are many such low-mass galaxies in the cluster, though most are without star-like nuclei.

In Fig. 8.43 we see a deeper, wider view of M100, which I made from the same AAT plates and in which more dwarf galaxies are visible. We might notice that beyond the clumpy, bright arms of the spiral, a smooth extension spreads faintly northwards in the direction of another dwarf elliptical, this one without a nucleus. This is about as faint as we can see with positive images from the plates I use for colour photography, but by combining the images from several exposures on high contrast plates from the UK Schmidt Telescope, we become aware that this small galaxy is caught up in the outskirts of M100 and has itself been distorted in the process, an effect clearly visible in Fig. 8.44. Moreover, this picture shows that the dwarf galaxy is itself involved with another, even fainter galaxy, just visible on the colour image. Might the influence of this insignificant pair of galaxies account for the asymmetry of M100? Possibly, but in the Virgo cluster there are many other candidates, and our understanding of interactions of this kind is limited by the lack of any sense of the third dimension, depth.

## CLUSTERS OF GALAXIES

As we leave the realm of individual galaxies and take in the ever wider view beyond a few tens of millions of light years, the photographs we obtain are less and less appealing to the eye, making the point that beyond galaxy-sized objects the Universe really is a pretty dark, empty place to our eyes. We can therefore offer few colour pictures of what are some of the largest single entities that we know of, the clusters of galaxies. The two closest to us naturally offer the best possibility of a photograph.

The enormous cloud of galaxies in Virgo is the nearest moderately rich cluster and is at a distance of about 50 million light years. Its members appear scattered over more than 100 square degrees of sky and the space between them looks empty, yet the individuals in the cluster are gravitationally bound one to the other. The region with the highest density of galaxies is that centred on the enormous ellipticals M84 and M86, which dominate Fig. 8.45, though the dynamical centre of the cluster is thought to be associated with the most massive member, the giant galaxy M87, pictured in Fig. 8.32. The most numerous of the more than 1000 members of the cluster are the much less conspicuous dwarf ellipticals. A few of these can be seen in this picture as faint fuzzy blobs, just vis-

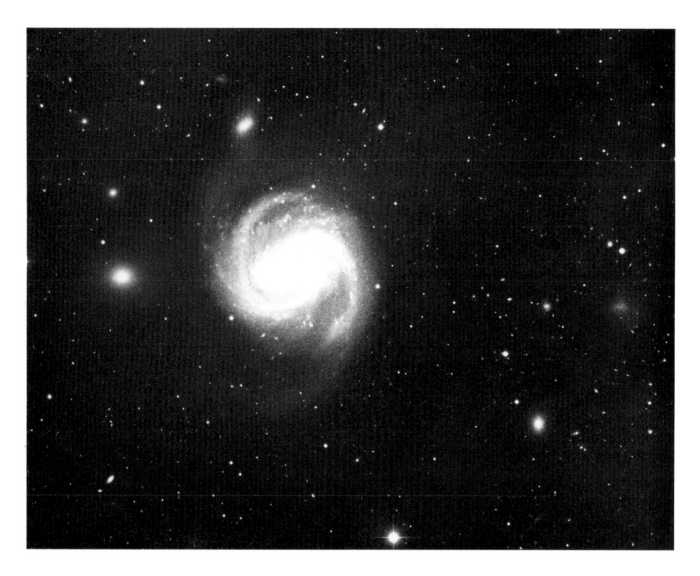

ible above the sky background in Fig. 8.45. More still are seen around M100, in the negative image of Fig. 8.44.

In common with all distant objects, and reflecting the expansion of the Universe, the whole Virgo cluster is receding with an average velocity of over 1000 kilometres per second. There is sufficient variation around that value to suggest that the cluster may be composed of more than one cloud of galaxies elongated along our line of sight. As we have remarked above, it is not surprising that close encounters between cluster members are fairly common in such a relatively crowded environment, and in the upper left (north-east) of Fig. 8.45, NGC 4438 has been distorted by its companion NGC 4435. On the very deepest pictures (Fig. 8.46) this pair displays a faint plume extending northwards and the faint outer envelopes of the two major galaxies in the colour picture seem to merge, though this is probably a perspective effect, making the point that these pictures are two-dimensional representations of three-dimensional structures.

Our second and last colour picture of a galaxy cluster is of that in the southern con-

*Fig. 8.43.* A deeper version of the previous picture shows many more of the faint dwarf galaxies in the Virgo cluster. Note especially the unusually blue but faint dwarf galaxy at the extreme right (west) of the picture and the faint extension of M100 northwards to the nucleated dwarf seen in Fig. 8.42.

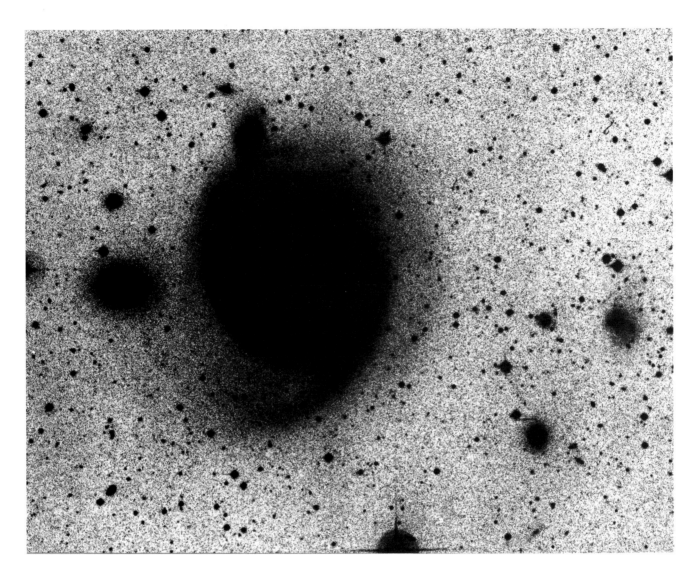

*Fig. 8.44.* A very deep negative print of the M100 region in Virgo, reproduced with the same scale and orientation as Fig. 8.43, shows the faintest galaxies. It is clear that the outer parts of M100 have been disturbed by interaction with faint dwarf galaxies.

stellation in Fornax. It is becoming evident that as we look to more distant objects the importance of colour pictures begins to diminish, which is why there are so few colour pictures of clusters of galaxies.

The galaxies in Fornax are a little more distant than those in Virgo and the field of view in Fig. 8.47 is considerably wider than that in Fig. 8.46. This was done to include the predominantly blue barred spiral NGC 1365 (lower right, better seen in Fig. 8.25) to compare with the much less interesting ellipticals, the biggest and brightest of which is NGC 1399. In common with many other relatively nearby ellipticals, I have searched NGC 1399 for shells and structures that might identify it as in some way optically peculiar, but without success.

As far as I can tell it is quite normal, but the photographs reveal something else. Figs. 8.48 (a) and (b) compare the same plate before and after copying with an unsharp mask. Apart from the swarm of globular clusters that this technique reveals, nothing exceptional is seen. However, a close look at the photograph shows a pair of interacting

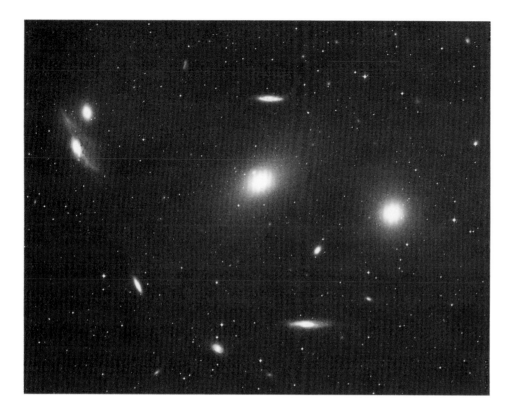

*Fig. 8.45.* M84 (right) and M86 are giant elliptical galaxies that lie at the optical centre of the Virgo cluster of galaxies. Several edge-on spirals and faint dwarfs can also be seen. The two galaxies at the upper left are an interacting pair, NGC 4435–38.

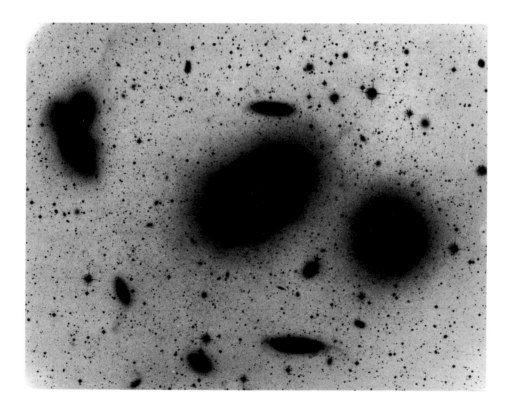

*Fig. 8.46.* A deeper, negative image of the centre of the Virgo cluster shows that the outer parts of M84 and M86 seem to overlap, though this is almost certainly a line-of-sight effect. The picture reveals many low-contrast anomalies in several of the bright galaxies in the field that are not evident on the colour picture.

galaxies far beyond the Fornax cluster, distant objects normally hidden by the bright envelope of NGC 1399. Even fainter and yet more distant galaxies are visible in the background, and the more carefully one looks the more galaxies are found as faint smudges all over the plate.

## THE MOST DISTANT VIEW

Our preoccupation with the spectacular subjects in this and previous chapters reflects our limited vision. Like the eye itself, the biggest telescopes tell us most about the nearest and brightest, and so those objects that make the best pictures tend to feature in this book. Again like the eye, when there are fewer photons we find it more difficult to discriminate objects by their colour. But that which is nearest and most conspicuous might not represent the Universe as a whole; indeed we know that it does not, for we have found that we can see only a small fraction of it, even that part which is nearby. Though the night sky seems filled with stars and simple optical aids reveal ever more of them, far beyond the stars of the Milky Way something else begins to be apparent.

When we turn the most powerful telescopes to those parts of the sky that are relatively free from the contamination of local points of light, we find fainter than the faintest stars are clouds of galaxies, so distant that they appear point-like, and almost impossible to distinguish from stars, even under the steadiest and darkest of skies. As we combine and manipulate more and more images and use ever more sensitive detectors to probe deeper and deeper we begin to realise that there are more of these distant galaxies than there are stars in the Milky Way, and yet we know from nearby examples that such systems may well contain as many or more stars as our home Galaxy.

Gradually we come to understand that our most distant view is filled with the tiny images of galaxies in groups, clusters and strings (Figs. 8.49, 8.50 and 8.51). Here and there in this seemingly endless faint mist are the bright images of quasars, enormously energetic galaxies that are often so distant that they look like bright stars, indistinguishable from those nearby by simple inspection of their images. Only their spectrum reveals their secret, and, paradoxically, their secret lies in their colour. These violently active galaxies are, for the most part, at high redshift, and the far ultraviolet parts of their spectra are shifted by the velocity of recession into the visible part of the spectrum. This does nothing for their visible colour and only the spectrograph can uncover this subtlety; in our pictures quasars look like stars.

In the realm of the quasars, we have left the nearby, extended objects that make spectacular pictures and we have thus reached the end of this photographic search for colour in the Universe. The images in this book are two-dimensional representations of three-dimensional space. Though we have manipulated the flat photographic negatives to give the impression of the third dimension, quantitative information about distance, and its equivalent, time, can only come indirectly, from photons sorted by means of a spectrograph. For all their astonishing revelations, spectrographs do not make pictures that people respond to, and since that is what this book is about, perhaps we should stop here. But before we do, one final point needs to be made.

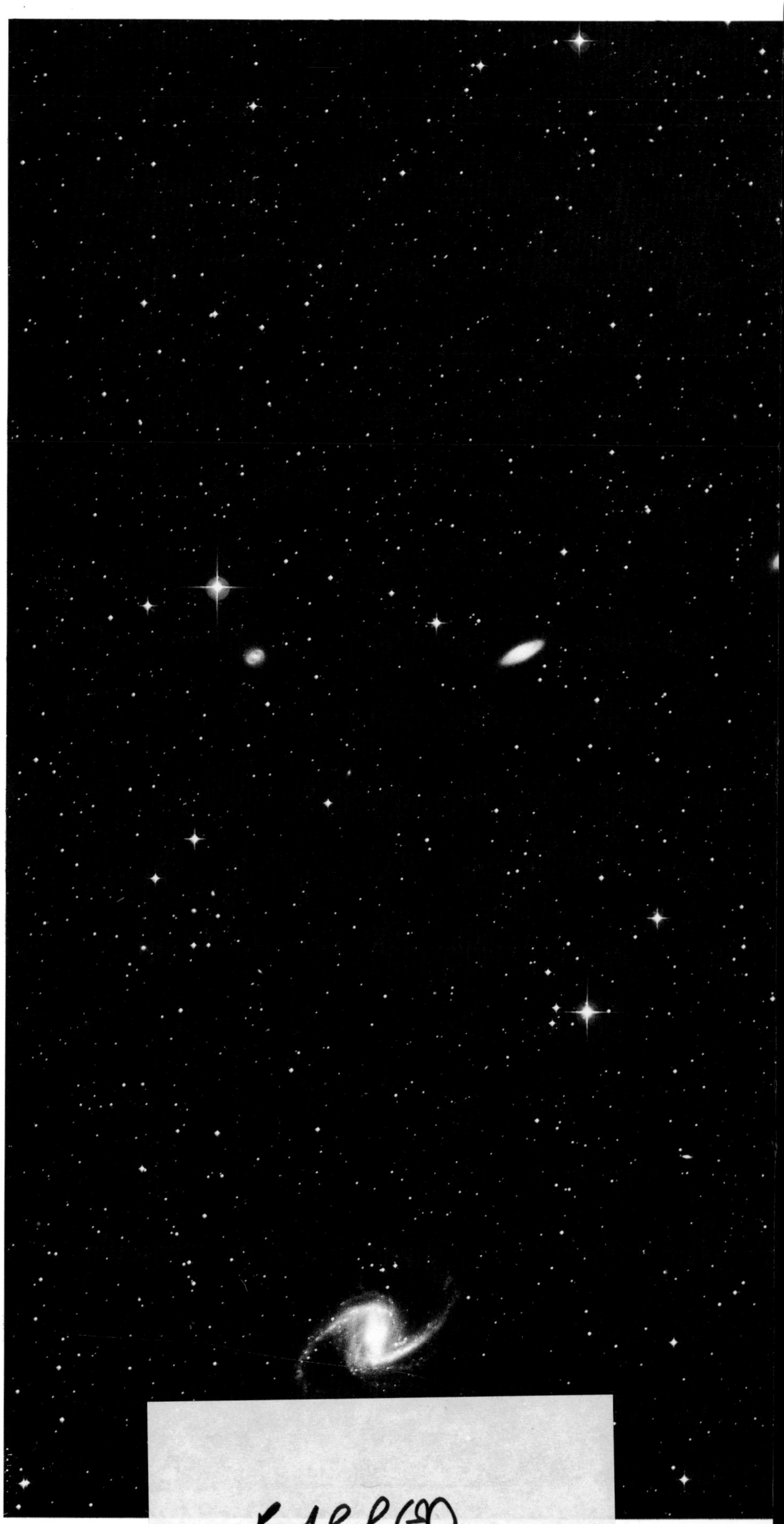

*Fig. 8.47.* Less rich in galaxies and more distant than the Virgo cluster is this group in the southern constellation of Fornax. The two bright ellipticals are NGC 1399 and NGC 1404. A contrast to the dominance of clusters by elliptical galaxies is the barred spiral NGC 1365, seen more clearly in Fig. 8.25.

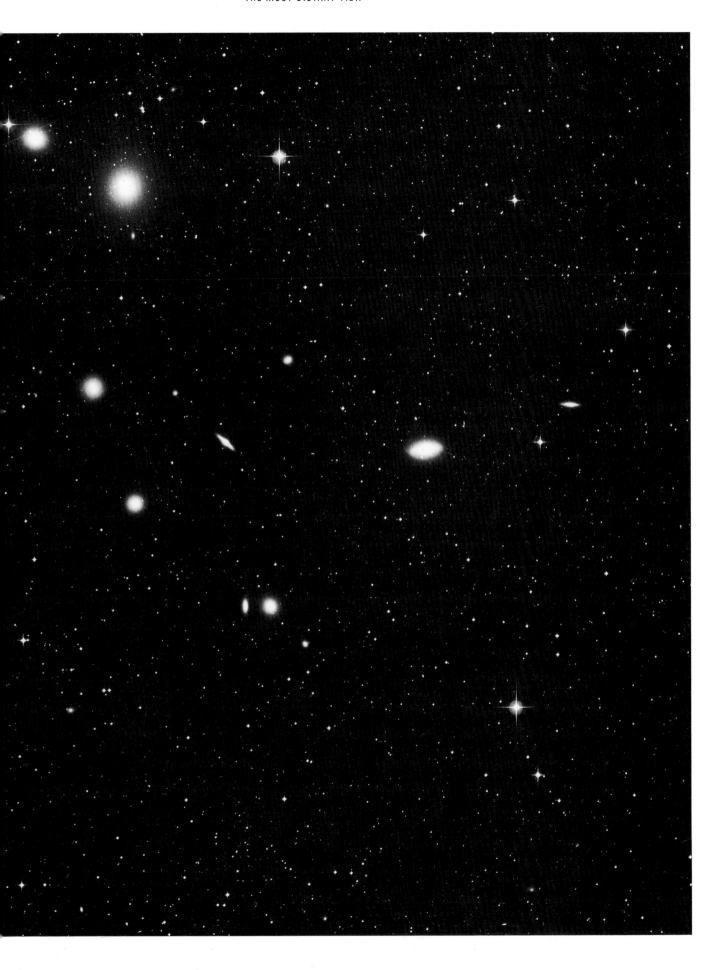

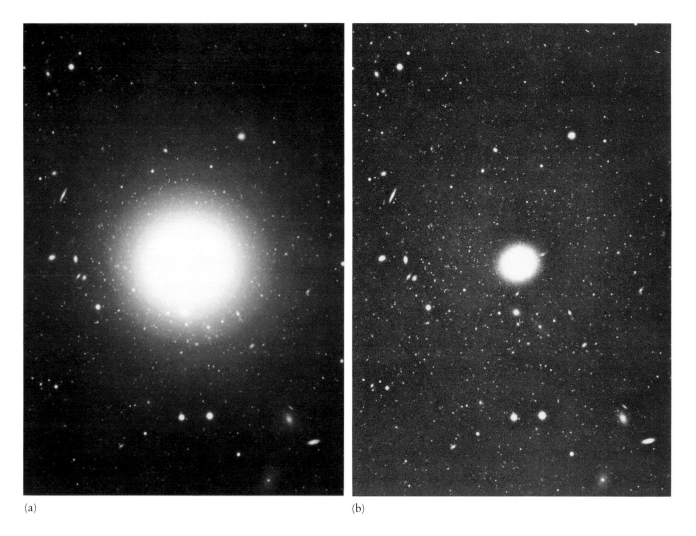

(a)                                                          (b)

## DARK MATTER

While acknowledging the limitations of colour photography, we also have to accept another, far more serious limitation. We find that we can only detect a small fraction of the Universe we know to be there.

There are curious similarities between the recent, gradual realisation that most of the Universe consists of some kind of 'dark matter' and William Herschel's long endeavour to uncover the 'construction of the heavens'. Herschel was not aware that his painstaking search, well under way 200 years ago, was seriously hampered by obscuring matter in the Milky Way, the dust that we have made so much of in this book. His view of the Universe was thus incomplete, but his work laid the foundations for understanding on which we still build. The reasons for the dark markings in the Milky Way and the 'zone of avoidance', where galaxies are not seen, were not firmly established until 100 years after Herschel's death. Of course, once it was appreciated that tiny, opaque particles were present as vast dark clouds between the stars, many other observational facts fell into place. Though it now seems quite obvious that we live in a dusty spiral galaxy, most of which is hidden from us by the dust itself, this fact was very difficult to

*Fig. 8.48.* Direct (a) and unsharp masked photographs (b) of the giant elliptical galaxy NGC 1399 in Fornax do not reveal any peculiarities, but the technique emphasises the cloud of globular clusters around the galaxy and shows numerous, faint galaxies in the background.

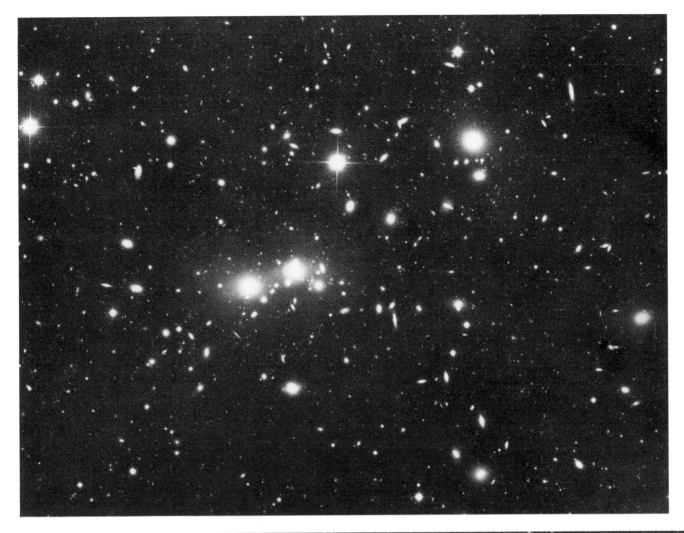

*Fig. 8.49.* A rich cluster of galaxies in Reticulum, known only by its catalogue name CA0340-538.

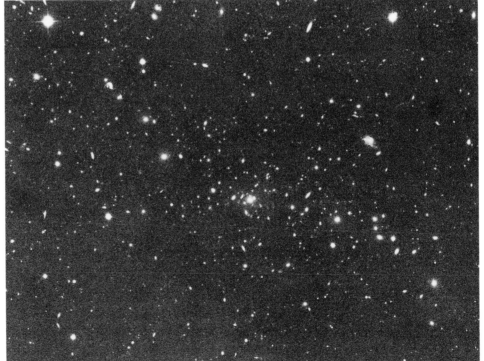

*Fig. 8.50.* The American astronomer George Abell catalogued thousands of distant clusters of galaxies from photographic plates. This one is Abell 222. Galaxies outnumber stars on this positive reproduction of a small part of an AAT plate.

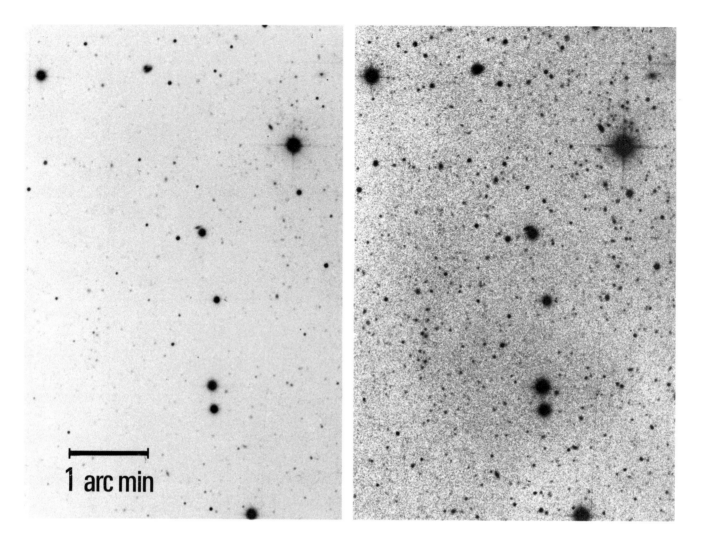

prove. The dust constitutes a minute part of the mass of a galaxy, but it has a profound effect on its appearance.

We are now faced with the same kind of dilemma, but now the unseen material is no minor constituent, it dominates the Universe. By measuring the velocities of galaxies around the centre of mass in large clusters, and the rotation of stars in the disks of spiral galaxies, it has become clear that both are strongly influenced by something that cannot otherwise be detected, the so-called dark matter. Other evidence for dark matter comes from the beautiful arcs that seem to embrace massive clusters of galaxies. These are a manifestation of a newly discovered phenomenon, the gravitational lens, which focuses distorted images of background galaxies in a way that seems to be associated with the foreground cluster. The arcs are well explained by the presence of huge amounts of unseen matter dominating the gravitational field within a cluster of galaxies. The size and location of the arcs can even be used to produce crude maps of the distribution of the dark matter within the group of galaxies, and estimates of the mass of the unseen component show that it completely dwarfs that due to the visible galaxies. Thus there is

*Fig. 8.51.* This pair of photographs are from the same original AAT plate. The image on the right has been enhanced to show the scatter of fuzzy blobs in the background. These are distant galaxies and on very deep photographs such as this they far outnumber the foreground stars. These galaxies are near the limit of what can be studied photographically from 4 metre-class, ground-based telescopes.

clearly a lot of it about, and several estimates suggest that dark matter comprises 90% of *all* matter, which means that our view of the Universe is remarkably incomplete, and we have only recently become aware of the problem.

Dark matter is no doubt telling us something of the origins of the Universe and will certainly determine its lifetime and its ultimate fate, yet we are left to guess what could have so much mass and be so widely distributed, yet remain invisible, undetectable at any wavelength. So, for all our enjoyment at discovering the colours of the luminous components in the night sky and our satisfaction in interpreting them, we still remain remarkably ignorant of the major constituent of the Universe.

# APPENDIX
# ABOUT THE PHOTOGRAPHS

## TECHNICAL DETAILS OF THE ASTRONOMICAL PHOTOGRAPHS

| Figure number | AAT or UKST ref. and/or title | Plate number(s) | Emulsion/filter (effective speed) | Exposure time | Observer/ photographer | Special process or camera |
|---|---|---|---|---|---|---|
| 1 | 2 | 3 | 4 | 5 | 6 | 7 |
| 3.3 | Sunset over the Warrumbungles | — | Ektachrome (400) ND7.0, then nil | 4 h, then 0.1 s | DFM | PP, Hasselblad 80 mm Planar |
| 3.5 | 1987 sunset | — | Ektachrome (400) | 5 s | DFM | PP, Hasselblad 80 mm Planar |
| 3.6 | January 1991 sunset | — | Ektachrome (400) | 5 s | DFM | PP, Hasselblad 80 mm Planar |
| 3.7 | The AAT dome in moonlight | — | Ektachrome (400) | 20 m | DFM | PP, Hasselblad 50 mm Distagon |
| 3.8 | Aurora Australis | — | Ektachrome (200) | 30 m | DFM | Hasselblad 50 mm Distagon |
| 3.9 | The zodiacal light | — | Ektachrome (400) | 20 m | DFM | PP, Hasselblad 80 mm Planar |
| 3.10 | AAT 46 Halley's comet | 2429 2428 2427 | IIa-O/GG 385 IIa-D/GG 495 098-04/RG 610 | 20 m 15 m 20 m | DFM _DAVID_ DFM _MALIN_ — DFM | |
| 3.11(a) 3.11(b) | Around the horizon at Siding Spring | — | 098-04 | 60 m | DFM | VM, Linhof 4 × 5 90 mm Rodagon |
| 3.12 | The lights of Punta Llana on La Palma | — | Fujichrome 400 | 5 h | DFM | Hasselblad 80 mm Planar |

| Figure number | AAT or UKST ref. and/or title | Plate number(s) | Emulsion/filter (effective speed) | Exposure time | Observer/ photographer | Special process or camera |
|---|---|---|---|---|---|---|
| 3.13 | The lights of Los Llanos on La Palma | — | Fujichrome 400 | 5 h | DFM | Hasselblad 50 mm Distagon |
| 3.14 | Stars rising over Coonabarabran | — | Ektachrome (400) | 2 h | DFM | PP, Hasselblad 80 mm Planar |
| 3.15 | Stars setting behind the AAT | — | Ektachrome (400) | 6 h | DFM | PP, Hasselblad 80 mm Planar |
| 3.16 | The south celestial pole | — | Ektachrome (400) | 9.5 h | DFM | PP, Hasselblad 50 mm Distagon |
| 3.17 | The south celestial pole | — | Ektachrome (400) | 3 h | DFM | PP, Hasselblad 50 mm Distagon |
| 3.18(a) | The north celestial pole | — | Fujichrome 400 | 6 h | DFM | Hasselblad 50 mm Distagon |
| 3.18(b) | The south celestial pole | — | Fujichrome 400 | 5 h | DFM | Hasselblad 50 mm Distagon |
| 3.19 | The northern horizon from Siding Spring | — | Ektachrome (400) | 6 h | DFM | PP, Hasselblad 50 mm Distagon |
| 3.20 | Star trails, Southern Cross and Pointers | — | 127-04 (IIIa-F) | 10.5 h | DFM | USM, Linhof 4 × 5 90 mm Rodagon |
| 4.2 | The stars of Orion, in focus | — | Ektachrome (400) | 45 m | DFM | PP, Hasselblad 80 mm Planar |
| 4.3 | The stars of Orion, out of focus | — | Ektachrome (400) | 45 m | DFM | PP, Hasselblad 80 mm Planar |
| 4.4 | Orion stars, step-focus | — | Ektachrome (400) | 30 m | DFM | PP, Hasselblad 80 mm Planar |
| 4.6 | Crux stars, step-focus star trail (1) | — | Ektachrome (400) | 35 m | DFM | PP, Hasselblad 80 mm Planar |
| 4.7 | Crux stars, step-focus star trail (2) | — | Ektachrome (400) | 45 m | DFM | PP, Hasselblad 80 mm Planar |
| 4.9 | The tail of the Scorpion | — | Ektachrome (400) | 40 m | DFM | PP, Hasselblad 80 mm Planar |

| Figure number | AAT or UKST ref. and/or title | Plate number(s) | Emulsion/filter (effective speed) | Exposure time | Observer/ photographer | Special process or camera |
|---|---|---|---|---|---|---|
| 4.10 | AAT 8, the spiral galaxy M83 | 2389 | IIa-O/GG 385 | 30 m | DFM | |
| | | 2390 | IIa-D/GG 495 | 30 m | DFM | — |
| | | 2391 | 098-04/RG 610 | 35 m | DFM | |
| 4.11 | Edge-on spiral galaxy NGC 891 | (INT)B89 | IIa-O/GG 385 | 35 m | The INT | |
| | | (INT)V60 | IIa-D/GG 495 | 40 m | Team | — |
| | | (INT)R90 | 098-04/RG 610 | 30 m | | |
| 4.12 | AAT 28, the Milky Way in Sagittarius | — | Ektachrome (400) | 90 m | DFM | PP, Hasselblad 80 mm Planar |
| 4.13 | UKS 10, NGC 6334 and NGC 6357 | B5129 | IIa-O/GG 385 | 50 m | SBT | |
| | | V5222 | IIa-D/GG 495 | 60 m | SBT | — |
| | | R5121 | 098-04/RG 610 | 60 m | SBT | |
| 4.15 | UKS 21, Baade's Window | B9503 | IIa-O/GG 385 | 30 m | MH | |
| | | V9459 | IIa-D/GG 495 | 30 m | MH | — |
| | | R9634 | 098-04/RG 630 | 60 m | MH | |
| 4.16 | UKS 20, Sagittarius star clouds | B9503 | IIa-O/GG 385 | 30 m | MH | |
| | | V9459 | IIa-D/GG 495 | 30 m | MH | — |
| | | R9524 | 098-04/RG 630 | 60 m | MH | |
| 4.18 | AAT 22, Barnard 86 and NGC 6520 | 1881 | IIa-O/GG 385 | 25 m | DFM | |
| | | 1882 | IIa-D/GG 495 | 25 m | DFM | — |
| | | 1883 | 098-04/RG 610 | 30 m | DFM | |
| 4.19 | AAT 25, NGC 4755, The Jewel Box | 2732 | IIa-O/GG 385 | 5 m | DFM | |
| | | 2733 | IIa-D/GG 495 | 5 m | DFM | — |
| | | 2734 | 098-04/RG 610 | 5 m | DFM | |
| 4.20 | Trumpler 5, an old open cluster | 2774 | IIa-O/GG 385 | 30 m | DFM | |
| | | 2775 | IIa-D/GG 495 | 25 m | DFM | — |
| | | 2776 | 098-04/RG 610 | 30 m | DFM | |
| 5.1 | UKS 22, M17 and NGC 6589–90 | B11167 | IIa-O/GG 385 | 50 m | MH | |
| | | V11168 | IIa-D/GG 495 | 50 m | MH | — |
| | | R11169 | 098-04/OG 630 | 60 m | MH | |
| 5.2 | AAT 20, NGC 6589–90 | 1895 | IIa-O/GG 385 | 35 m | DFM | |
| | | 1894 | IIa-D/GG 495 | 30 m | DFM | — |
| | | 1893 | 098-04/RG 610 | 40 m | DFM | |

| Figure number | AAT or UKST ref. and/or title | Plate number(s) | Emulsion/filter (effective speed) | Exposure time | Observer/ photographer | Special process or camera |
|---|---|---|---|---|---|---|
| 5.3 | UKS 1, the Horsehead nebula | B5443 | IIa-O/GG 385 | 60 m | JB | |
| | | X5448 | IIa-D/GG 445 | 60 m | SBT | — |
| | | R5437 | 098-04/RG 610 | 60 m | SBT | |
| 5.4 | Part of IC 434 | 2290 | 098/RG 610 | 35 m | DFM | USM |
| 5.5 | AAT 36, the Horsehead nebula | 2292 | IIa-O/GG 385 | 25 m | DFM | |
| | | 2291 | IIa-D/GG 495 | 25 m | DFM | — |
| | | 2290 | 098-04/RG 610 | 35 m | DFM | |
| 5.6 | Herbig–Haro nebulae | 2290 | 098-04/RG 610 | 35 m | DFM | USM |
| 5.7 | AAT 74, reflection nebula in NGC 6188 | 2544 | IIa-O/GG 385 | 30 m | DFM | |
| | | 2543 | IIa-D/GG 495 | 25 m | DFM | — |
| | | 2542 | 098-04/RG 610 | 35 m | DFM | |
| 5.8a 5.8b | UKS 11 and 12, NGC 6334 and NGC 6357 | B5129 | IIa-O/GG 385 | 50 m | SBT | |
| | | V5222 | IIa-D/GG 495 | 60 m | SBT | — |
| | | R5121 | 098-04/RG 610 | 60 m | SBT | |
| 5.9 | AAT 38, NGC 3576 and NGC 3603 | 2306 | IIa-O/GG 385 | 20 m | DFM | |
| | | 2307 | IIa-D/GG 495 | 25 m | DFM | — |
| | | 2308 | 098-04/RG 610 | 35 m | DFM | |
| 5.10 | Wide-angle view of Crux and Carina | — | Ektachrome (400) | 45 m | DFM | PP, Hasselblad 80 mm Planar |
| 5.12 | AAT 10, open cluster NGC 3923 | 2732 | IIa-O/GG 385 | 5 m | DFM | |
| | | 2733 | IIa-D/GG 495 | 5 m | DFM | — |
| | | 2734 | 098-04/RG 610 | 5 m | DFM | |
| 5.13 | UKS 18, the Pleiades cluster | B8960 | IIa-O/GG 385 | 30 m | MH | |
| | | V8826 | IIa-D/GG 495 | 30 m | MH | PA |
| | | R8935 | 098-04/RG 630 | 40 m | MH | |
| 5.14 | UKS 4, Rho Ophiuchi and Antares | B5183 | IIa-O/GG 385 | 60 m | JB | |
| | | V5120 | IIa-D/GG 495 | 60 m | JB | PA |
| | | R5128 | 098-04/RG 630 | 60 m | SBT | |
| 5.15 | AAT 24, IC 2220, Toby Jug nebula | 1807 | IIa-O/GG 385 | 40 m | DFM | |
| | | 1806 | IIa-D/GG 495 | 35 m | DFM | — |
| | | 1805 | 098-04/RG 630 | 35 m | DFM | |
| 5.16 | Rho Ophiuchi nebula | J645 | IIIa J/GG 395 | 50 m | PC | USM |

| Figure number | AAT or UKST ref. and/or title | Plate number(s) | Emulsion/filter (effective speed) | Exposure time | Observer/ photographer | Special process or camera |
|---|---|---|---|---|---|---|
| 5.17 | Orion nebula reflected in NGC 1977 | B5443 | IIa-O/GG 385 | 60 m | JB | |
| | | V4664 | IIa-D/GG 495 | 60 m | SBT | USM, PA |
| | | R5437 | 098-04/RG 630 | 60 m | SBT | |
| 5.18 | AAT 34, NGC 1973–75–77 | 2189 | IIa-O/GG 385 | 15 m | DFM | |
| | | 2188 | IIa-D/GG 495 | 15 m | DFM | — |
| | | 2187 | 098-04/RG 610 | 15 m | DFM | |
| 5.20 | AAT 32, Eta Carinae nebula | 2361 | IIa-O/GG 385 | 15 m | DFM | |
| | | 2362 | IIa-D/GG 495 | 15 m | DFM | USM |
| | | 2363 | 098-04/RG 610 | 15 m | DFM | |
| 5.21 | AAT 66, SN1987A light echo | 2591—2311 | IIa-O/GG 385 | 35/20 m | DFM/DFM | |
| | | 2590—2312 | IIa-D/GG 495 | 30/25 m | DFM/DFM | USM, PA, SUB |
| | | 2592—2310 | 098-04/RG 610 | 40/35 m | DFM/DFM | |
| 5.24 | UKS 7, the M20 reflection nebula | B5130 | IIa-O/GG 385 | 40 m | SBT | |
| | | V5223 | IIa-D/GG 495 | 60 m | SBT | USM, PA |
| | | R4977 | 098-04/RG 630 | 60 m | JB | |
| 5.25 5.26 | Parts of the Trifid nebula | 2622 | IIa-O/GG 385 | 35 m | DLJ | |
| | | 2623 | IIa-D/GG 495 | 30 m | DLJ | USM, PA |
| | | 2624 | 098-04/RG 610 | 30 m | DLJ | |
| 5.27 | AAT 71, the CG 4 cometary globule | 2718 | IIa-O/GG 385 | 35 m | DH | |
| | | 2724 | IIa-D/GG 495 | 30 m | DH | PA |
| | | 2725 | 098-04/RG 630 | 35 m | DH | |
| 5.28 | Cometary globule tail | J3915 | IIIa-J/GG 395 | 64 m | JAD | PA |
| 5.29 | AAT 72, Sco dark cloud | 2750 | IIa-O/GG 385 | 19 m | DFM | |
| | | 2751 | IIa-D/GG 495 | 25 m | SL | PA |
| | | 2758 | 098-04/RG 630 | 35 m | SL | |
| 5.30 | AAT 73, Corona Australis nebula | 2759 | IIa-O/GG 385 | 25 m | DFM | |
| | | 2760 | IIa-D/GG 495 | 25 m | DFM | — |
| | | 2761 | 098-04/RG 610 | 35 m | DFM | |
| 5.31 | Corona Australis | UKST plates | 5 × 098-04 | 5 × 30 m | — | SUPER |
| 6.1 | AAT 12, M20, the Trifid nebula | 2622 | IIa-O/GG 385 | 35 m | DLJ | |
| | | 2623 | IIa-D/GG 495 | 30 m | DLJ | — |
| | | 2624 | 098-04/OG 610 | 30 m | DLJ | |

| Figure number | AAT or UKST ref. and/or title | Plate number(s) | Emulsion/filter (effective speed) | Exposure time | Observer/ photographer | Special process or camera |
|---|---|---|---|---|---|---|
| 6.2 | AAT 69, faint nebula around M20 | 2622 | IIa-O/GG 385 | 35 m | DLJ | |
| | | 2623 | IIa-D/GG 495 | 30 m | DLJ | USM, PA |
| | | 2624 | 098-04/OG 610 | 30 m | DLJ | |
| 6.3 | AAT 30, stars in M20 | 2555 | IIa-O/GG 385 | 3 m | DFM | |
| | | 2556 | IIa-D/GG 495 | 3 m | DFM | — |
| | | 2467 | 098-04/RG 610 | 3.5 m | DFM | |
| 6.4 | AAT 31 the Hourglass nebula | 2465 | IIa-O/GG 385 | 3 m | DFM | |
| | | 2466 | IIa-D/GG 495 | 2.5 m | DFM | USM |
| | | 2467 | 098-04/RG 610 | 3.5 m | DFM | |
| 6.5 | UKS 5, M8, the Lagoon nebula | B5130 | IIa-O/GG 385 | 40 m | SBT | |
| | | V5223 | IIa-D/GG 495 | 60 m | SBT | USM |
| | | R4977 | 098-04/RG 630 | 60 m | JB | |
| 6.6 | The region around M8 and M20 | B5130 | IIa-O/GG 385 | 40 m | SBT | |
| | | V5223 | IIa-D/GG 495 | 60 m | SBT | — |
| | | R4977 | 098-04/RG 630 | 60 m | JB | |
| 6.7 | UKS 3, NGC 6559 and IC 1274–75 | B5130 | IIa-O/GG 385 | 40 m | SBT | |
| | | V5223 | IIa-D/GG 495 | 60 m | SBT | — |
| | | R4977 | 098-04/RG 630 | 60 m | JB | |
| 6.8 | UKS 23, wide-angle view of Orion nebulae | B5443 | IIa-O/GG 385 | 60 m | JB | |
| | | V5448 | IIa-D/GG 495 | 60 m | SBT | — |
| | | R5437 | 098-04/RG 630 | 60 m | SBT | |
| 6.9 | Lynds 1630 nebulosity | R1151 | 098-04/RG 630 | 60 m | MES | USM, PA |
| 6.10 | Trapezium stars | AAT F/8 | Plus-X/nil | 1 m | DFM | USM |
| 6.11 | AAT 29, at the heart of the Orion nebula | 2714 | IIa-O/GG 385 | 1 m | DFM | |
| | | 2715 | IIa-D/GG 495 | 1 m | DFM | USM |
| | | 2716 | 098-04/RG 630 | 1.2 m | DFM | |
| 6.12 | AAT 19, the Orion nebula | 1651 | IIa-O/GG 385 | 5 m | DFM | |
| | | 1650 | IIa-D/GG 495 | 5 m | DFM | USM |
| | | 1649 | 098-04/RG 630 | 5 m | DFM | |
| 6.13 | UKS 8, deep image of the Orion nebula | B5443 | IIa-O/GG 385 | 60 m | JB | |
| | | V5448 | IIa-D/GG 495 | 60 m | SBT | USM |
| | | R5437 | 098-04/RG 630 | 60 m | SBT | |

| Figure number | AAT or UKST ref. and/or title | Plate number(s) | Emulsion/filter (effective speed) | Exposure time | Observer/ photographer | Special process or camera |
|---|---|---|---|---|---|---|
| 6.14 | UKS 9, the Rosette nebula | B6692 | IIa-O/GG 385 | 60 m | DHM | — |
| | | V5493 | IIa-D/GG 495 | 60 m | MH | |
| | | R7515 | 098-04/RG 630 | 60 m | DHM | |
| 6.15 | AAT 35, the centre of the Rosette nebula | 1798 | IIa-O/GG 385 | 35 m | DFM | — |
| | | 1799 | IIa-D/GG 475 | 30 m | DFM | |
| | | 1800 | 098-04/RG 630 | 40 m | DFM | |
| 6.16 | IC 1340, part of the Cygnus Loop | (INT)B84 | IIa-O/GG 385 | 37 m | The INT | — |
| | | (INT)V83 | IIa-D/GG 495 | 37 m | Team | |
| | | (INT)R85 | 098-04/RG 610 | 30 m | | |
| 6.17 | AAT 14, nebulosity near S Monocerotis | 1774 | IIa-O/GG 385 | 30 m | DFM | — |
| | | 1775 | IIa-D/GG 495 | 30 m | DFM | |
| | | 2302 | IIIa F/RG 630 | 90 m | DFM | |
| 6.18 | AAT 13, the Cone nebula | 1774 | IIa-O/GG 385 | 30 m | DFM | |
| | | 1775 | IIa-D/GG 495 | 30 m | DFM | USM |
| | | 2302 | IIIa F/RG 630 | 90 m | DFM | |
| 6.19 | The Cone nebula and Trumpler 5 | B6692 | IIa-O/GG 385 | 60 m | DHM | — |
| | | V5493 | IIa-D/GG 495 | 60 m | MH | |
| | | R7515 | 098-04/RG 630 | 60 m | DHM | |
| 6.20 | AAT 18, the M16 nebula | 2485 | IIa-O/GG 385 | 18 m | RMS | |
| | | 2486 | IIa-D/GG 495 | 18 m | RMS | USM |
| | | 2487 | 098-04/RG 610 | 24 m | RMS | |
| 6.21 | AAT 47, dust lanes in M16 | 2485 | IIa-O/GG 385 | 18 m | RMS | |
| | | 2486 | IIa-D/GG 495 | 18 m | RMS | USM |
| | | 2487 | 098-04/OG 610 | 24 m | RMS | |
| 6.22 | Bok globules in NGC 3676 | 2306 | IIa-O/GG 385 | 20 m | WDP | |
| | | 2307 | IIa-D/GG 495 | 25 m | DFM | — |
| | | 2308 | 098-04/RG 610 | 35 m | DFM | |
| 6.23 | AAT 77, Bok globules in IC 2944 | 2742 | IIa-O/GG 385 | 25 m | SL | |
| | | 2741 | IIa-D/GG 495 | 25 m | SL | — |
| | | 2743 | 098-04/RG 610 | 35 m | DFM | |
| 6.24 | UKS 6, the Eta Carinae nebula | B2210 | IIa-O/GG 385 | 15 m | MES | |
| | | V2209 | IIa-D/GG495 | 15 m | MES | USM |
| | | R2211 | 098-04/RG 630 | 20 m | MES | |

| Figure number | AAT or UKST ref. and/or title | Plate number(s) | Emulsion/filter (effective speed) | Exposure time | Observer/ photographer | Special process or camera |
|---|---|---|---|---|---|---|
| 6.25 | AAT 9, the Eta Carinae nebula | 2361 | IIa-O/GG 385 | 15 m | DFM | — |
|  |  | 2362 | IIa-D/GG 495 | 15 m | DFM |  |
|  |  | 2363 | 098-04/OG 610 | 15 m | DFM |  |
| 6.26 | AAT 37, Eta Carinae and Trumpler 14 | 2361 | IIa-O/GG 385 | 15 m | DFM | USM |
|  |  | 2362 | IIa-D/GG 495 | 15 m | DFM |  |
|  |  | 2363 | 098-04/RG 610 | 15 m | DFM |  |
| 6.27 | AAT 40, the NGC 3603 cluster | 2306 | IIa-O/GG 385 | 20 m | WDP | — |
|  |  | 2307 | IIa-D/GG 495 | 25 m | DFM |  |
|  |  | 2308 | 098-04/RG 610 | 35 m | DFM |  |
| 6.28 | Around the Tarantula nebula | B2179 | IIa-O/GG 385 | 20 m | AJL | PA |
|  |  | V2169 | IIa-D/GG 495 | 20 m | AJL |  |
|  |  | R2178 | 098-04/RG 630 | 30 m | AJL |  |
| 6.30 | AAT 44, the Tarantula nebula | 2311 | IIa-O/GG 385 | 20 m | DFM | — |
|  |  | 2312 | IIa-D/GG 495 | 25 m | DFM |  |
|  |  | 2310 | 098-04/RG 610 | 35 m | DFM |  |
| 6.31 | AAT 68, stars around 30 Doradus | 2714 | IIa-O/GG 385 | 1 m | DFM | — |
|  |  | 2715 | IIa-D/GG 495 | 1 m | DFM |  |
|  |  | 2716 | 098-04/OG 610 | 1.25 m | DFM |  |
| 7.2 | Herbig–Haro objects | 2290 | 098-04/RG 610 | 35 m | DFM | USM |
| 7.3 | A Herbig–Haro nebula in NGC 2264 | 1774 | IIa-O/GG 385 | 30 m | DFM | USM |
|  |  | 1775 | IIa-D/GG 495 | 30 m | DFM |  |
|  |  | 2302 | IIIa F/RG 630 | 90 m | DFM |  |
| 7.4 | AAT 45, the Homunculus nebula | — | Vericolor film, no filter | 1 m (AAT/f8) | DFM | — |
| 7.5 | Faint wisps near Eta Carinae | 2361 | IIa-O/GG 385 | 15 m | DFM | USM |
|  |  | 2362 | IIa-D/GG 495 | 15 m | DFM |  |
|  |  | 2363 | 098-04/RG 610 | 15 m | DFM |  |
| 7.6 | AAT 33, Henize 70 nebula in the LMC | 2190 | IIa-O/GG 385 | 25 m | DFM | PA |
|  |  | 2192 | IIa-D/GG 497 | 25 m | DFM |  |
|  |  | 2191 | 098-04/RG 610 | 25 m | DFM |  |
| 7.7 | Bubble nebulae in NGC 6822 | 2054 | IIa-O/GG 385 | 30 m | DFM | PA |
|  |  | 2053 | IIa-D/GG 495 | 30 m | DFM |  |
|  |  | 2052 | 098-04/RG 630 | 40 m | DFM |  |

| Figure number | AAT or UKST ref. and/or title | Plate number(s) | Emulsion/filter (effective speed) | Exposure time | Observer/ photographer | Special process or camera |
|---|---|---|---|---|---|---|
| 7.8 | Barnard's Loop | — | IIIa F/RG 610 | 120 m | DFM | 180 mm Aero-Ektar PA, VC |
| 7.9 | Bubble nebulae in NGC 6822 | 2054 | IIa-O/GG 385 | 30 m | DFM | |
| | | 2053 | IIa-D/GG 495 | 30 m | DFM | PA |
| | | 2052 | 098-04/RG 630 | 40 m | DFM | |
| 7.10 | AAT 16, Wolf–Rayet star in NGC 2359 | 1669 | IIa-O/GG 385 | 35 m | DFM | |
| | | 1668 | IIa-D/GG 495 | 35 m | DFM | — |
| | | 1667 | 098-04/RG 630 | 35 m | DFM | |
| 7.11 | NGC 7635, the Bubble nebula | (INT)B71 | IIa-O/GG 385 | 40 m | The INT | |
| | | (INT)V70 | IIa-D/GG 495 | 25 m | Team | — |
| | | (INT)R56 | 098-04/RG 610 | 36 m | | |
| 7.12 | AAT 21, the bipolar nebula NGC 6164–65 | 1878 | IIa-O/GG 385 | 30 m | DFM | |
| | | 1879 | IIa-D/GG 475 | 30 m | DFM | — |
| | | 1880 | 098-04/RG 630 | 35 m | DFM | |
| 7.13 | AAT 27, around NGC 6164–65 nebula | 1878 | IIa-O/GG 385 | 30 m | DFM | |
| | | 1879 | IIa-D/GG 475 | 30 m | DFM | PA |
| | | 1880 | 098-04/RG 630 | 35 m | DFM | |
| 7.14 | The nebula around Antares | B5183 | IIa-O/GG 385 | 60 m | JB | |
| | | V5120 | IIa-D/GG 495 | 60 m | SBT | PA, USM |
| | | R5128 | 098-04/RG 630 | 60 m | SBT | |
| 7.15 | AAT 24, Toby Jug nebula, deep picture | 1807 | IIa-O/GG 385 | 40 m | DFM | |
| | | 1806 | IIa-D/GG 495 | 35 m | DFM | PA |
| | | 1805 | 098-04/RG 630 | 35 m | DFM | |
| 7.16 | AAT 15, The Helix nebula, NGC 7293 | 1725 | IIa-O/GG 385 | 35 m | DFM *DAVID* | PA |
| | | 1724 | IIa-D/GG 495 | 35 m | DFM *MALIN* | PA |
| | | 1636 | IIIa F/RG 630 | 90 m | DFM | USM |
| 7.17 | The Dumbbell nebula, NGC 6583, M27 | (INT)B68 | IIa-O/GG 385 | 35 m | The INT | |
| | | (INT)V67 | IIa-D/GG 495 | 20 m | Team | — |
| | | (INT)R93 | 098-04/RG 610 | 25 m | | |
| 7.18 | AAT 81, Ack 277-03.1 planetary nebula | 2740 | IIa-O/GG 385 | 30 m | DFM | |
| | | 2729 | IIa-D/GG 495 | 30 m | DFM | PA |
| | | 2730 | 098-04/RG 630 | 35 m | DFM | |

| Figure number | AAT or UKST ref. and/or title | Plate number(s) | Emulsion/filter (effective speed) | Exposure time | Observer/ photographer | Special process or camera |
|---|---|---|---|---|---|---|
| 7.19 | AAT 79, NGC 2346, a planetary nebula | 2738 | IIa-O/GG 385 | 20 m | DFM | |
| | | 2739 | IIa-D/GG 495 | 20 m | DFM | — |
| | | 2511 | 098-04/RG 630 | 30 m | DFM | |
| 7.20 | AAT 80, NGC 2899, a planetary nebula | 2740 | IIa-O/GG 385 | 30 m | DFM | |
| | | 2729 | IIa-D/GG 495 | 30 m | DFM | — |
| | | 2730 | 098-04/RG 630 | 30 m | DFM | |
| 7.21 | AAT 11, NGC 6302, planetary nebula | — | Vericolor negative/GG385 | 45 m | DFM | — |
| 7.22 | AAT 48b, SN1987A + precursor | 2521 | IIa-O/GG 385 | 20 m | RMS | Plate 2310 over-laid as a negative (see Fig. 6.30) |
| | | 2522 | IIa-D/GG 495 | 15 m | RMS | |
| | | 2523 | 098-04/RG 610 | 20 m | RMS | |
| 7.23 | AAT 67, SN1987A after four years | 2714 | IIa-O/GG 385 | 1 m | DFM | |
| | | 2715 | IIa-D/GG 495 | 1 m | DFM | |
| | | 2716 | 098-04/RG 610 | 1.25 m | DFM | |
| 7.24 | The Crab nebula | Z387 | 103a O/nil | 10 m | Not known | (Hale 200″ plates) USM |
| | | Z368 | 103a D/GG 11 | 10 m | | |
| | | Z376 | 103a E/'Chance red' | 20 m | | |
| 7.25 | The Crab nebula polarisation structure | 5129 | 103a-D/4384 | 20 m | Not recorded | (Hale 200″ plates) SUB |
| | | 5130 | 103a-D/4384 | 20 m | | |
| 7.26 | UKS 2, the Vela supernova remnant | B4703 | IIa-O/GG 385 | 60 m | MH | |
| | | V4671 | IIa-D/GG 495 | 60 m | PRS | PA |
| | | R4679 | 098-04/RG 630 | 60 m | PRS | |
| 7.27 | AAT 78, part of the Vela SNR | 2787 | IIIa-J/Bj | 101 m | DFM | |
| | | 2773 | IIIa-F/Vf | 100m | DFM | PA |
| | | 2772 | IIIa-F/RG 610 | 101 m | DFM | |
| 7.28 | NGC 6995, part of the Cygnus Loop | (INT)B84 | IIa-O/GG 385 | 37 m | The INT Team | |
| | | (INT)V83 | IIa-D/GG 495 | 37 m | | — |
| | | (INT)R85 | 098-04/RG 610 | 30 m | | |
| 7.29 | AAT 76, 47 Tucanae globular cluster | 1995 | IIa-O/GG 385 | 10 m | DFM | |
| | | 1994 | IIa-D/GG 495 | 10 m | DFM | USM |
| | | 2765 | 098-04/RG 610 | 12 m | DFM | |

| Figure number | AAT or UKST ref. and/or title | Plate number(s) | Emulsion/filter (effective speed) | Exposure time | Observer/ photographer | Special process or camera |
|---|---|---|---|---|---|---|
| 7.30 | AAT 70, NGC 5904, M5 cluster | 2752 | IIIa-J/Bj | 2 m | SL | |
| | | 2753 | IIIa-F/Vf | 2 m | SL | — |
| | | 2754 | IIIa-F/RG 610 | 2 m | SL | |
| 7.31 | AAT 42, Hodge 11, a globular cluster | 2018 | IIa-O/GG 385 | 10 m | DFM | |
| | | 2019 | IIa-D/GG 495 | 10 m | DFM | — |
| | | 2020 | 098-04/RG 630 | 15 m | DFM | |
| 8.1 | The Magellanic Clouds | — | Fujichrome 400 | 90 m | DFM | Hasselblad 80 mm Planar |
| 8.2 | AAT 41, NGC 4027 | 2083 | IIa-O/GG 385 | 15 m | DAH | |
| | | 2084 | IIa-D/GG 495 | 15 m | DAH | — |
| | | 2085 | 098-04/RG 630 | 20 m | DAH | |
| 8.3 | The LMC | — | 3 × Fujichrome 400 frames | 3 × 90 m | DFM | SUPER |
| 8.4 | UKS 14, the LMC | B1141 | IIa-O/GG 385 | 40 m | MES | |
| | | V1884 | IIa-D/GG 495 | 60 m | MES | — |
| | | R1151 | 098-04/RG 630 | 60 m | MES | |
| 8.5 | UKS 16, star clouds north of the LMC | B1141 | IIa-O/GG 385 | 40 m | MES | |
| | | V1884 | IIa-D/GG 495 | 60 m | MES | — |
| | | R1151 | 098-04/RG 630 | 60 m | MES | |
| 8.6 | Henize 55, LMC bubble nebula | B1141 | IIa-O/GG 385 | 40 m | MES | |
| | | V1884 | IIa-D/GG 495 | 60 m | MES | — |
| | | R1151 | 098-04/RG 630 | 60 m | MES | |
| 8.7 | UKS 15, nebulae around 30 Doradus | B2179 | IIa-O/GG 385 | 20 m | AJL | |
| | | V2169 | IIa-D/GG 495 | 20 m | AJL | PA |
| | | R2178 | 098-04/RG 630 | 30 m | AJL | |
| 8.8 | UKS 17, the SMC | B1876 | IIa-O/GG 385 | 55 m | MES | |
| | | V1840 | IIa-D/GG 495 | 40 m | MES | — |
| | | R998 | 098-04/RG 630 | 60 m | DJF | |
| 8.9 | AAT 51, The Leo 1 dwarf galaxy | 2537 | IIa-O/GG 385 | 25 m | DFM | |
| | | 2536 | IIa-D/GG 495 | 20 m | DFM | PA |
| | | 2388 | 098-04/RG 610 | 30 m | DFM | |
| 8.10 | The Fornax dwarf | 2182 | IIIa-J | 80 m | DAH | PA |

| Figure number | AAT or UKST ref. and/or title | Plate number(s) | Emulsion/filter (effective speed) | Exposure time | Observer/ photographer | Special process or camera |
|---|---|---|---|---|---|---|
| 8.11 | The Sextans dwarf | J10147 | IIIa-J/GG 395 | 65 m | MH | |
| | | J10658 | IIIa-J/GG 395 | 80 m | AS | SUPER |
| | | J12465 | IIIa-J/GG 395 | 120 m | PM | |
| 8.12 | AAT 26, NGC 6822 in the Local Group | 2054 | IIa-O/GG 385 | 30 m | DFM | |
| | | 2053 | IIa-D/GG 495 | 30 m | DFM | PA |
| | | 2052 | 098-04/RG 630 | 40 m | DFM | |
| 8.13 | AAT 43, IC 5152 | 2119 | IIa-O/GG 385 | 15 m | Information | |
| | | 2120 | IIa-D/GG 495 | 15 m | not | — |
| | | 2121 | 098-04/RG 630 | 20 m | avaiable | |
| 8.14 | M31 (Palomar 48″ Schmidt) | — | Super Ancochrome | 120 m | WCM | — |
| 8.15 | NGC 205 | KP3141 | IIIa-F/RG 610 | 60 m | Carney | USM |
| 8.16 | The bulge of M31 | (INT)B97 | IIa-O/GG 385 | 20 m | The INT | Vignetting |
| | | (INT)V98 | IIa-D/GG 495 | 25 m | Team | mask |
| | | (INT)R99 | 098-04/RG 610 | 20m | | |
| 8.17 | The nucleus of M31 | (INT)B97 | IIa-O/GG 385 | 20 m | The INT | |
| | | (INT)V98 | IIa-D/GG 495 | 25 m | Team | — |
| | | (INT)R99 | 098-04/RG 610 | 20 m | | |
| 8.18 | AAT 23, edge-on spiral, NGC 253 | 1040 | IIa-O/GG 385 | 25 m | KCF | |
| | | 1041 | IIa-D/GG 495 | 25 m | KCF | USM |
| | | 1039 | 098-04/RG 630 | 25 m | KCF | |
| 8.19 | AAT 56, NGC 300 | 2581 | IIa-O/GG 385 | 40 m | RMS | |
| | | 1889 | IIa-D/GG 495 | 35 m | DFM | PA |
| | | 2704 | 098-04/RG 610 | 40 m | DFM | |
| 8.20 | AAT 57, the nucleus of NGC 300 | 2581 | IIa-O/GG 385 | 40 m | RMS | |
| | | 1889 | IIa-D/GG 495 | 35 m | DFM | — |
| | | 2704 | 098-04/RG 610 | 40 m | DFM | |
| 8.22 | M33 | (INT)B100 | IIa-O/GG 385 | 35 m | The INT | |
| | | (INT)V101 | IIa-D/GG 495 | 40 m | Team | PA |
| | | (INT)R102 | 098-04/RG 610 | 35 m | | |
| 8.22 | The inner regions of M33 | (INT)B100 | IIa-O/GG 385 | 35 m | The INT | |
| | | (INT)V101 | IIa-D/GG 495 | 40 m | Team | — |
| | | (INT)R102 | 098-04/RG 610 | 35 m | | |

| Figure number | AAT or UKST ref. and/or title | Plate number(s) | Emulsion/filter (effective speed) | Exposure time | Observer/ photographer | Special process or camera |
|---|---|---|---|---|---|---|
| 8.23 | AAT 8, M83 | 2389 | IIa-O/GG 385 | 30 m | DFM | |
| | | 2390 | IIa-D/GG 495 | 30 m | DFM | — |
| | | 2391 | 098-04/RG 610 | 35 m | DFM | |
| 8.24 | AAT 17, spiral galaxy NGC 2997 | 1782 | IIa-O/GG 385 | 30 m | DC | |
| | | 1718 | IIa-D/GG 495 | 25 m | DC | — |
| | | 1784 | 098-04/RG 610 | 40 m | DC | |
| 8.25 | AAT 55, barred spiral, NGC 1365 | 2720 | IIa-O/GG 385 | 30 m | DFM | |
| | | 2721 | IIa-D/GG 495 | 30 m | DFM | — |
| | | 2722 | 098-04/RG 610 | 40 m | DFM | |
| 8.26 8.27 | AAT 54, Seyfert galaxy, NGC 1566 | 2300 | IIa-O/GG 385 | 25 m | WDP | |
| | | 2301 | IIa-D/GG 495 | 25 m | WDP | PA, SUP |
| | | 2299 | 098-04/RG 630 | 35 m | WDP | |
| 8.28 | AAT 7, NGC 5128, Centaurus A | 2472 | IIIa-J/Bj | 85 m | DFM | |
| | | 2462 | IIIa-F/Vf | 90 m | RMS | — |
| | | 2541 | IIIa-F/RG 610 | 80 m | DFM | |
| 8.29 | AAT 52, the dust lane of Centaurus A | 2472 | IIIa-J/Bj | 85 m | DFM | |
| | | 2462 | IIIa-F/Vf | 90 m | RMS | USM |
| | | 2541 | IIIa-F/RG 610 | 80 m | DFM | |
| 8.30 | Centaurus A shells | CTIO 4m | IIIa-J | 55 m | John Graham | USM |
| 8.31 | NGC 1344 shells | UKST plates | 3 x IIIa-J/GG 395 | 200 m total | — | SUPER |
| 8.32 | AAT 60, NGC 4486, (M87). Jet inset | 2540 | IIa-O/GG 385 | 30 m | RDC | USM (inset only) |
| | | 2539 | IIa-D/GG 495 | 25 m | RDC | |
| | | 2566 | 098-04/RG 630 | 50 m | DFM | |
| 8.33 | AAT 53, the M87 globular clusters | 2540 | IIa-O/GG 385 | 30 m | RDC | |
| | | 2539 | IIa-D/GG 495 | 25 m | RDC | USM |
| | | 2566 | 098-04/RG 630 | 30 m | DFM | |
| 8.34 | Malin-1 | UKST plates | 4 × IIIa-J/GG 395 | 224 m total | — | PA, SUPER |
| 8.35 | NGC 4038–39, the Antennae | J3968 | IIIa-J/GG 395 | 70 m | JAD | PA |
| 8.36 | AAT 65, NGC 4038–39 inner part | 2744 | IIa-O/GG 385 | 25 m | DFM | |
| | | 2745 | IIa-D/GG 495 | 25 m | DFM | — |
| | | 2746 | 098-04/RG 610 | 35 m | DFM | |

| Figure number | AAT or UKST ref. and/or title | Plate number(s) | Emulsion/filter (effective speed) | Exposure time | Observer/ photographer | Special process or camera |
|---|---|---|---|---|---|---|
| 8.37 | AAT 85, the group of galaxies in Leo | 2595 | IIa-O/GG 385 | 27 m | DFM | |
| | | 2596 | IIa-D/GG 495 | 22 m | DFM | — |
| | | 2597 | 098-04/RG 610 | 27 m | DFM | |
| 8.38 | AAT 61, M65 (NGC 3623) in Leo | 2595 | IIa-O/GG 385 | 27 m | DFM | |
| | | 2596 | IIa-D/GG 495 | 22 m | DFM | — |
| | | 2597 | 098-04/RG 610 | 27 m | DFM | |
| 8.39 | AAT 62, M66 (NGC 3627) in Leo | 2595 | IIa-O/GG 385 | 27 m | DFM | |
| | | 2596 | IIa-D/GG 495 | 22 m | DFM | — |
| | | 2597 | 098-04/RG 610 | 27 m | DFM | |
| 8.40 | AAT 63, NGC 3628 in the Leo Group | 2595 | IIa-O/GG 385 | 27 m | DFM | |
| | | 2596 | IIa-D/GG 495 | 22 m | DFM | — |
| | | 2597 | 098-04/RG 610 | 27 m | DFM | |
| 8.41 | AAT 64, Starburst galaxy NGC 1313 | 2768 | IIa-O/GG 385 | 30 m | DFM | |
| | | 2767 | IIa-D/GG 495 | 25 m | DFM | PA |
| | | 2766 | 098-04/RG 610 | 30 m | DFM | |
| 8.42 8.43 | AAT 58, NGC 4321, M100 | 2744 | IIa-O/GG 385 | 25 m | DFM | PA |
| | | 2745 | IIa-D/GG 495 | 25 m | DFM | (8.42 only) |
| | | 2746 | 098-04/RG 610 | 35 m | DFM | |
| 8.44 | Faint dwarf galaxies around M100 | UKST plates | 4 × IIIa-J/GG 395 | Approx. 250 m | — | PA, SUPER |
| 8.45 | UKS 24, the centre of the Virgo cluster | B9263 | IIa-O/GG 385 | 30 m | JB | |
| | | V9262 | IIa-D/GG 495 | 30 m | JB | PA |
| | | R9266 | 098-04/RG 630 | 30 m | JB | |
| 8.46 | Negative of Virgo cluster centre | 6 × UKST IIIa-J/GG 395 exposures | | 450 m total | — | SUPER |
| 8.47 | UKS 13, the Fornax cluster | B2694 | IIa-O/GG 385 | 40 m | PRS | |
| | | V2695 | IIa-D/GG 495 | 50 m | PRS | — |
| | | R4711 | 098-04/RG 630 | 60 m | MH | |
| 8.48 | NGC 1399 | 1556 | IIIa-J/GG 385 | 90 m | Not recorded | USM (b only) |
| 8.49 | CA 0340–538 cluster | 2059 | IIIa-J/GG 385 | 90 m | BLW | PA |
| 8.50 | The Abell 222 cluster | 1763 | IIIa-J/GG 385 | 70 m | DC | PA |
| 8.51 | Distant cluster | 1566 | IIIa-J/GG 385 | 90 m | BLW | PA |

## NOTES TO THE APPENDIX

**Column 2.** A brief description of the photograph and the AAT or UKS reference number. In other publications the prefix UKS may appear as 'ROE'. Both prefixes refer to 24 colour photographs made with plates from the UK Schmidt Telescope (UKST) when it was operated by the Royal Observatory, Edinburgh (ROE). The prefix AAT refers to the Anglo-Australian Telescope.

**Column 3.** Plate numbers are given for all the three-colour pictures. Those prefixed by capital letters are UK Schmidt exposures and follow the UKST practice of identifying the pass-band with a single letter, usually B (blue), V ('visual' or green-yellow) and R (red). Where many similar plates contribute to a single black and white image they are not identified individually, but the total exposure time is reported in Column 5. A few photographs made with plates from the Isaac Newton Telescope have been used throughout and these are prefixed 'INT' followed by a letter that identifies the pass-band in the manner of the UKST plates. Plate numbers without prefixes are always AAT exposures. KP3141 (Fig. 8.15) is a plate from the Kitt Peak 4 metre telescope.

**Column 4** records the combinations of plates and filters used to make the three-colour illustrations or the film type used for direct colour photography. In many cases standard transparency films have been push processed (PP) for extra speed and contrast, and in these cases the effective nominal ISO film speed is given in parentheses.

Other emulsion types listed are usually Eastman Kodak spectroscopic emulsions, though a few other commercially available materials such as Eastman Kodak's Plus-X black and white film and Vericolor negative film are also recorded. All spectroscopic emulsions exposed on the AAT are hypersensitised before use by baking for a few hours in nitrogen at 65 °C followed by a soak in pure hydrogen at 20 °C, again for a few hours, the exact times being strongly batch-dependent and determined afresh for each hypering run. At the UKST, the fast, relatively coarse-grained type IIa and 098-04 materials are not normally hypersensitised; however, fine-grain blue-green sensitive IIIa-J and its red-sensitive counterpart IIIa-F are treated before exposure by a prolonged (several weeks) nitrogen soak at 20 °C followed by a few hours in hydrogen at 20 °C.

**Column 6** lists by their initials the many people who contributed their time and expertise to taking plates that have been used for the illustrations in this book. Some have helped me with the AAT service photography programme or other photographic projects on the AAT and many have been staff members at the UKST, where it is the custom to take plates on behalf of others. The superb quality of these plates has made a major contribution to this book. It is a pleasure to identify all concerned.

## THE PEOPLE WHO TOOK THE PLATES

| | | | | |
|---|---|---|---|---|
| JB | John Barrow | WDP | Bill Pence | |
| RDC | Russell Cannon | AS | Ann Savage | |
| DC | David Carter | RMS | Ray Sharples | |
| PC | Peter Corben | MES | Liz Sim | |
| JAD | John Dawe | PRS | Peter Standen | |
| DJF | David Finch | SBT | Sue Tritton | |
| KCF | Ken Freeman | BLW | B. Louise Webster | |
| MH | Malcolm Hartley | | | |
| DH | Despina Hatzidimitriou | | | |
| DAH | David Hanes | | | |
| DLJ | Dave Jauncey | | | |
| SL | Steve Lee | | | Miguel Briganti |
| AJL | Andy Longmore | | | Lius Chinnaro |
| DFM | David Malin | The Isaac Newton | | Derek Jones |
| PM | Trish McKenzie | Telescope (INT) Team | | Luis Manade |
| WCM | Bill Miller | | | Gabriel Perez |
| DHM | David Morgan | | | Pepe Vilchez |

**Column 7** indicates which of several possible techniques have been incorporated into the photograph. In this table these techniques are identified by abbreviations as follows:

PP – Push processing
USM – Unsharp masking
SUB – Image subtraction
VM – Vignetting masking
PA – Photographic amplification
SUPER – superimposition

Further details of these and other relevent photographic processes are given in the Bibliography.

# BIBLIOGRAPHY

The photographic processes referred to throughout this book are generally described in a non-technical way that gives the flavour of the effect they have on the final photograph. However, some of the effects of image manipulation are quite subtle and to appreciate these reference to my original publications is encouraged. Of course, this is also necessary if anyone should wish to try or modify any of the processes I have used to make the astronomical pictures that appear here.

The techniques have been developed over a long period and have resulted in many publications. Some of the techniques were specifically devised for particular astronomical photographic problems, and are described in the discovery papers, while other publications are of more general nature or were directed to photographic rather than astronomical readers. The following is a list of my main scientific publications with any photographic content, together with a selection of publications showing the range of useful things that can be done in astronomy with photography. I have also included a few 'popular' articles that explore topics dealt with elsewhere in this book. In most cases the titles of these are self explanatory.

## UNSHARP MASKING

**Unsharp masking.** *AAS Photo-Bull.* No.16, 10–13 (1977).
> This is the definitive reference for application of unsharp masking in astronomy. Full details of the process are given, including establishing the sensitometric and spatial properties of the mask itself. Though these principles have not changed, I no longer use the processing recommendations given here for making unsharp masks, but the preferred film remains the same.

**Astrophotography with unsharp masking** (with W.J. Zealey). *Sky and Telescope* **57**, 354–359, April, 1979.
**Unsharp masking** *J. Astron. Soc. of Pacific* (Mercury) 8 No. 4, 89–93 (1979).
> Two general articles on the topic.

**New optical observations of galactic supernova remnants** (with W.J. Zealey and K.H. Elliott). *Astron. and Astrophys. Supp.* **38** 39–49 (1979).

**The structure of the faint nebulosity and obscuration toward M8, M20 and W28** (with H. Hartl, H.T. MacGillivray and W.J. Zealey). *Ap. J. Lett.* **23** 193–205 (1983).

> Photographic amplification and unsharp masking used together to reveal complex, low contrast but sharp-edged supernova remnants and other structures in a crowded star field containing diffuse nebulosity.

**Direct photographic image enhancement in astronomy.** *J. Phot. Sci.* **29** No. 5, 199–205 (1981).

> A review of photographic amplification and unsharp masking from the photographic science point of view. The paper also includes a demonstration of image superimposition for improving the signal to noise ratio by combining several black and white images.

**Shell structure in NGC 5128** (with P.J. Quinn and J.H. Graham). *Ap. J. Lett.* **272** L5–L7 (1983).

> The shells in Centaurus A were discovered because this plate was being used to demonstrate the technique of unsharp masking to the CTIO photographers.

**Structures in the gas tail of Comet Halley.** *Nature* **320** 577 (1986).

## PHOTOGRAPHIC AMPLIFICATION

**Photographic amplification of faint astronomical images.** *Nature* **276** 591–593 (1978).

> The first publication of the technique of photographic amplification discusses the basic principle behind the method and describes the process itself.

**A jet associated with M89.** *Nature* **277** 279–280 (1979).

> Photographic amplification and unsharp masking used together to reveal faint and low contrast features associated with an elliptical galaxy. Though not appreciated at the time, M89 was the first 'shell' galaxy.

**Giant shells around elliptical galaxies** (with D. Carter). *Nature* **285** 643–645 (1980).

> The discovery of faint shells around apparently normal elliptical galaxies.

**Photographic enhancement of direct astronomical images.** *AAS Photo-Bull.* No. 27, No. 2, 4–9 (1981).

> Reviews of photographic techniques with a more detailed discussion of the mechanism of photographic amplification, illustrated with many examples.

**A catalog of elliptical galaxies with shells** (with D. Carter). *Ap. J.* **274** 534–540 (1983).

> While the discovery of shells around elliptical galaxies was unexpected, the generality of the phenomenon was very surprising. This paper lists 150 new examples and illustrates some unusual specimens, mostly discovered using photographic amplification on UK Schmidt plates.

**Faint object detection on photographic plates** (with W.J. Couch, R.S. Ellis, E.J. Kibblewhite and J. Godwin). *MNRAS* **209** 307–315 (1984).

> A comparison of the detection of faint images of clusters of galaxies using the eye and automatic measuring machines on photographically amplified images.

**Faint structures in the NGC 2261 region** (with J.R. Walsh). *MNRAS* **217** 31–40 (1985).

> A study of Herbig–Haro objects associated with Hubble's variable nebula, triggered by photographic amplification of a photographic plate of the region.

**Food for photometrists – faint galaxies revealed**. In *New Aspects of Galaxy Photometry*, ed. J-L. Nieto, Lecture Notes in Physics series, No. 232 (Springer-Verlag 1985), pp.27–32.

**Infrared cirrus and HI in directions near the LMC** (with R.X. McGee, R.F. Hanes and R.J.-M. Grognard). *MNRAS* **221** 543–552 (1986).

> The detection of very faint, large-scale galactic reflection nebulosity easily confused with the Magellanic Stream.

**Herbig–Haro objects in the vicinity of NGC 2023** (with K. Ogura and J.R. Walsh). *MNRAS* **227** 361–372 (1987).

> The discovery of Herbig–Haro objects at the base of the Horsehead nebula.

**Discovery of a huge, low surface brightness galaxy: a proto-disk galaxy at low redshift?** (with G.D. Bothun, C.D. Impey and J.R. Mould). *A.J.* **94** 23–29 (1987).

> The discovery of Malin-1 during a search of UK Schmidt plates for faint galaxies in the Virgo cluster.

**The AAO distant galaxy cluster survey** (with W.J. Couch and R.S. Ellis). In *The World of Galaxies*, ed. Corwin and Bottinelli. Proceedings of Conference Le Monde des Galaxies, Paris, April 1988. (Springer-Verlag, New York, 1989), pp. 25–28.

**A uniformly-selected catalogue of distant galaxy clusters** (with W.J. Couch, R.S. Ellis and I. MacLaren). *MNRAS* **249** 606–628 (1991).

> Photographic amplification used to select images of very distant clusters of galaxies recorded serendipitously on AAT photographic plates taken for other purposes.

**What are the emission line filaments along the radio axis of Centaurus A?** (with R. Morganti, C.N. Tadhunter and R.A.E Fosbury). In *BL Lac Objects*, ed. L. Maraschi, T. Maccacaro and M-H. Ulrich. Lecture Notes in Physics series (Springer-Verlag, Berlin, 1989).

**The nature of the optical filaments in Centaurus A: evidence for a beamed ionising continuum** (with R. Morganti, A. Robinson, R.A.E. Fosbury, S. di Serego Alighieri and C.N. Tadhunter). *MNRAS* **249** 91–112 (1991)

> Faint filaments in the outer envelope of the galaxy are examined and considered as evidence that from certain perspectives, the relativistic beam from the active nucleus of the galaxy can be 'seen' by material at considerable distances from the source. The filaments themselves are quite insignificant, and require image enhancement to make them visible, but are evidence that Centaurus A is the nearest quasar-like galaxy.

**An optical counterpart to the H I cloud in the Local Supercluster** (with C.D. Impey, G. Bothun and L. Staveley-Smith). *Ap. J. Lett.* **351** L33–L35 (1990).

> The discovery of a large cloud of hydrogen gas relatively nearby led to the search for a star-forming region within it. This paper reports the discovery of a faint, irregular galaxy in just the right place.

## COMBINED TECHNIQUES, IMAGE ADDITION AND SUBTRACTION

**The interaction between the relativistic jets of SS433 and the interstellar medium** (with W.J. Zealey and M.A. Dopita). *MNRAS* **92** 731–743 (1980).

> The discovery of faint wisps of nebulosity associated with the famous jets of SS433.

**Virgo dwarfs: new light on faint galaxies** (with G.D. Bothun and C.D. Impey). *Ap. J.* **330** 634–660 (1988).

> A follow-up study of a large number of very faint galaxies in the Virgo cluster, initially selected from UK Schmidt plates by photographic amplification and image addition.

**The superimposition of multiple plates.** *Astrophotography*, Proceedings of 1987 IAU Photographic Working Group meeting, Jena, April 1987, ed. S. Marx. (Springer-Verlag, Heidelberg, 1988), pp.125–132.

>    A demonstration of the value of adding together photographically amplified copies of large numbers of plates. An image obtained by combining 36 plates is reproduced.

**Photographic imaging of light echoes from SN 1987A** (with W.J. Couch and D.A. Allen). *MNRAS* **242** 555–559 (1989).

>    The first publication to show clearly the complexity of the light echoes of supernova 1987A, revealed by a photographic amplification and subtraction technique.

**The extended red emission filaments of NGC 2023** (with A.N. Witt). *Ap. J. Lett.* **347** L25–L27 (1989).

>    The discovery of filaments of anomalous red emission in the well-known reflection nebula at the base of the Horsehead nebula by subtracting images of two colours after unsharp masking.

**Extracting the image in astronomical photography.** *TechBits Magazine*, (Eastman Kodak) Issue 1, 1–10 (1990).

>    A well-illustrated, largely non-technical description of the black and white photographic processes that are useful in astronomy, especially as a preliminary to making colour pictures. Part 2 of this article (listed under Colour Photography) shows how the black and whites are made into colour images.

## HYPERSENSITISATION

**The effect of environment on the sensitivity of hydrogen hypersensitised plates.** *Modern Techniques in Astronomical Photography*. Proceedings of ESO Conference, Geneva, ed. West and Heudier (ESO, 1978), pp. 107–112.

>    Hypersensitising a dried emulsion with hydrogen is an important part of astronomical photography for both amateur and professional. It is not widely appreciated that the dried emulsion rapidly absorbs moisture from its surroundings during exposure, negating the beneficial effects of hypersensitisation. This paper was an early attempt to quantify that loss of long-exposure speed.

## COLOUR PHOTOGRAPHY

**Colour photography on the 3.9 m AAT and 1.2 m Schmidt Telescopes.** *Modern Techniques in Astronomical Photography*. Proceedings of ESO Conference, Geneva, ed. West and Heudier (ESO, 1978), pp. 235–237.

>    Early results using colour film and a brief discussion of the reasons for not using it.

**Colour photography in astronomy.** *Vistas in Astronomy* **24** Pt 3, 219–238 (1980).

>    The first comprehensive technical account of three-colour photography in astronomy, including a review of the origins of colour photography and earlier work in astronomy with colour film.

**Colours of the stars** (with Paul Murdin) (Cambridge University Press, 1984), pp. 198.

>    The Technical Appendix of this book summarises the three-colour process of making astronomical colour photographs.

**Astronomical color photography.** *McGraw-Hill Encyclopedia of Science and Technology*, 7th edn (McGraw-Hill, Maidenhead, 1985).

>    Encyclopaedia article on colour photography in astronomy.

**A universe of color.** *TechBits Magazine* (Eastman Kodak) Issue 3, 3–11 (1990).
> A largely non-technical, well-illustrated description of three-colour photography in astronomy. The article compares the spectral response of colour film, Tech Pan and spectroscopic plates.

**A celebration of colour in astrophysics.** *Current Science* (Indian Academy of Sciences) **60** No. 1 (1991) 59pp (complete issue).
> The publication that grew into this book.

## GENERAL SCIENTIFIC ARTICLES AND REVIEWS

**Special photographic techniques applied to Schmidt plates.** Proceedings of the Second IAU Asian-Pacific Regional meeting, Bandung, Indonesia, 1981, ed. B. Hidayat, (Tira Pustaka, Jakarta, 1984) pp. 398–401.

**Photographic image intensification and reduction: a unified optical approach.** *J. Phot. Sci.* **30** No. 3, 87–94 (1982).
> The diffuse-light contact copying process used for photographic amplification can also be used to recover information from seriously under- or over-exposed conventional negatives. This paper demonstrates a direct, non-destructive substitute for chemical intensification or reduction using a direct duplication film designed for graphic arts applications. The paper also includes a novel way of controlling the contrast of direct duplicating materials over a very wide range using a pre-flashing technique.

**The detection of faint objects against the sky background.** *Astronomy with Schmidt-type telescopes.* IAU Colloquium No. 78, Asiago, Aug–Sept 1983, ed. M. Capaccioli (D. Reidel, Dordrecht, 1984) pp. 57–72.

**Where to next in astronomical photography?** *Proceedings of the 1984 IAU Photographic Working Group Meeting*, Edinburgh, 3–6 April, 1984, ed. M.E. Sim and H. Ishida (Royal Observatory, Edinburgh, 1984) pp. 111–118.
> Many examples of discoveries made using recently-developed photographic techniques and a discussion of new materials for astronomical photography in its future prospects.

**Advanced techniques for astronomical photography and additive colour photography in astronomy.** *Temmon Gaido*, Tokyo, December 1986 (in Japanese) pp. 133–148.

**The age of the specialist in astronomical photography.** *Astrophotography*, Proceedings of 1987 IAU Photographic Working Group meeting, Jena, April 1987, ed. S. Marx (Springer-Verlag, Heidelberg, 1988), pp. 2–20.
> A discussion of the modernisation of the way professional astronomical photography is practised in the electronic age.

**Astronomical photography under the microscope.** *Today's Science Tomorrow's Technology*, ed. M.H. Brennan and D.D. Millar (Pergamon Press, Sydney 1989) pp. 87–104.
> A historical review of the gradual invention of photography and of the contribution astronomers have made to the improvement of the photographic process since its discovery.

**The astronomical photography of Gerard de Vaucouleurs.** In *A Life for Astronomy*, ed. M. Capaccioli and H.G. Corwin (World Scientific Publishing Co., Singapore, 1989), pp. 53–64.
> A review which spans 45 years, covering the photographic work of one of the best-known astronomers. De Vaucouleurs' efforts to make photography a quantitative tool has led to many discoveries, especially in the field of galaxies.

## 'POPULAR' ARTICLES

The deep sky in color. *Sky and Telescope* **62** 216–219, September 1981.

The Orion Nebula in color. *Sky and Telescope* **62** 414–417, November 1981.

A look at some unstable stars. *Sky and Telescope* **63** 22–26, January 1981.

Improved techniques for astrophotography. *Sky and Telescope* **62** 4–7, July 1981.

The dust clouds of Sagittarius. *Sky and Telescope* **63** 254–259, March 1982.

Searching the southern skies. *Australian Natural Hist.* **20** No. 9, 285–289 (1982).

Shooting stars. Qantas *Inflight* Magazine, 6–10, March/April 1986.

In search of star colors. *Sky and Telescope* **72** 326–330, October 1986.

The splendor of Eta Carinae. *Sky and Telescope* **73** 14–18, January 1987.

More about star colours. (Letter to the Editor) *Sky and Telescope* **74** 358, April 1987.

In the shadow of the Horsehead Nebula. *Sky and Telescope* **74** 253–257, September 1987.

Things to see and do in the dark. *J. Brit. Astron. Assoc.* **97** 288–290, August 1987.

The developing art of star photography. *New Scientist* **120** 23–28, 17 December 17, 1988.

NGC 3576 and 3603 in Carina. *Sky and Telescope* **78** 144–145, August 1989.

Star trails and quantum leaping: Astronomy for new audiences. (audio-visual production with Martin Wesley-Smith) In *Newsletter* of the Australian Astronomy and Space Liason Group, ed. Nick Lomb, No. 15, 51–53, March 1989.

Deep space photography. *Guide to Australian Astronomy*, Federal Publishing Co., 56–67, July 1989.

Echoes of the supernova. (With David Allen) *Sky and Telescope* **79** 22–25, January 1990.

Peculiar galaxies I have known. *Southern Astronomy* **2** 20–27 (1990).

Photographing the colours of the stars. *Australian Geographic*, Issue 19, p. 103, July 1990.

Beeldbewerking in de astrophotographie. *Zenit* (Utrecht) 1/91, 4–13.

Des geschiedenis van de astrofotografie. *Zenit* (Utrecht) 1/92, 4–11.

Astronomy in colour. *Astronomy Now*, May 1992, p. 7.

Kleurrijk heelal. *Zenit* (Utrecht) 5/92, 196–203.

## JOURNAL ABBREVIATIONS

| | |
|---|---|
| *AAS Photo-Bull.* | American Astronomical Society Photo-Bulletin |
| *A.J.* | Astronomical Journal |
| *Ap. J.* | Astrophysical Journal |
| *Ap. J. Lett.* | Astrophysical Journal (Letters) |
| *Astron. and Astrophys. Supp.* | Astronomy and Astrophysics Supplement |
| *J. Phot. Sci.* | Journal of Photographic Science |
| *MNRAS* | Monthly Notices of the Royal Astronomical Society |

# Index

Page numbers in bold type refer to figures